LIFE ON T[...]

LEVEL 1   LEVEL [...]   LEVEL 4

DRINKING WATER

TRANSPORTATION

COOKING

EATING

SLEEPING

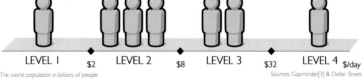

LEVEL 1 $2   LEVEL 2 $8   LEVEL 3 $32   LEVEL 4 $/day

The world population in billions of people   Sources: Gapminder[3] & Dollar Street

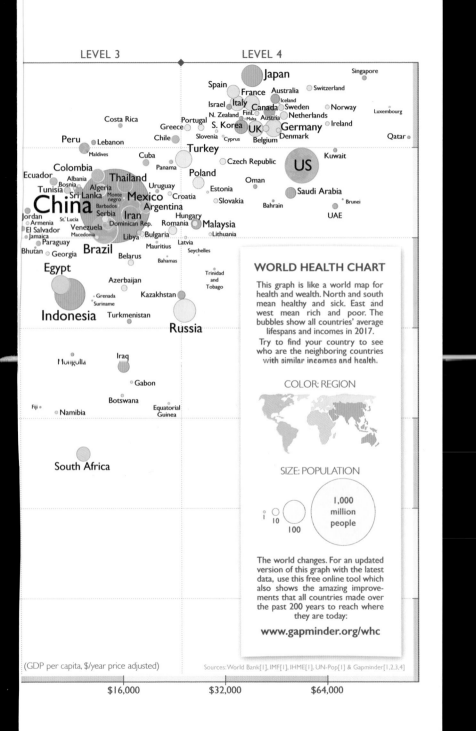

LEVEL 3   LEVEL 4

Singapore

Japan

Spain
France   Australia   Switzerland
Israel   Italy   Canada   Iceland
N. Zealand   Finl.   Sweden   Norway   Luxembourg
Costa Rica   Portugal   Malta   Austria   Netherlands
Greece   S. Korea   UK   Germany   Ireland
Chile   Slovenia   Cyprus   Belgium   Denmark   Qatar
Peru   Lebanon
Maldives   Cuba   Turkey
Colombia   Panama   Czech Republic   Kuwait
Ecuador   Albania   Poland   US
Tunisia   Bosnia   Algeria   Thailand   Uruguay   Oman
Sri Lanka   Monte   Mexico   Croatia   Estonia   Saudi Arabia
China   negro   Argentina   Slovakia   Bahrain   Brunei
Jordan   Barbados   Serbia   Iran   Dominican Rep.   Romania   Malaysia   UAE
Armenia   St. Lucia   Venezuela   Libya   Bulgaria   Lithuania
El Salvador   Macedonia
Jamaica   Paraguay   Latvia
Bhutan   Georgia   Brazil   Mauritius   Seychelles
Egypt   Belarus   Bahamas
Trinidad
Azerbaijan   and
Tobago
Grenada   Kazakhstan
Suriname
Indonesia   Turkmenistan
Russia

Mongolia   Iraq

Gabon

Botswana
Fiji   Equatorial
Namibia   Guinea

South Africa

WORLD HEALTH CHART

This graph is like a world map for
health and wealth. North and south
mean healthy and sick. East and
west mean rich and poor. The
bubbles show all countries' average
lifespans and incomes in 2017.

Try to find your country to see
who are the neighboring countries
with similar incomes and health.

COLOR: REGION

SIZE: POPULATION

1,000
million
people

1
10
100

The world changes. For an updated
version of this graph with the latest
data, use this free online tool which
also shows the amazing improvements that all countries made over
the past 200 years to reach where
they are today:

www.gapminder.org/whc

(GDP per capita, $/year price adjusted)

Sources: World Bank[1], IMF[1], IHME[1], UN-Pop[1] & Gapminder[1,2,3,4]

$16,000 $32,000 $64,000

PEOPLE BY REGION AND INCOME

2017

Number of people in different regions

Each figure is one billion people

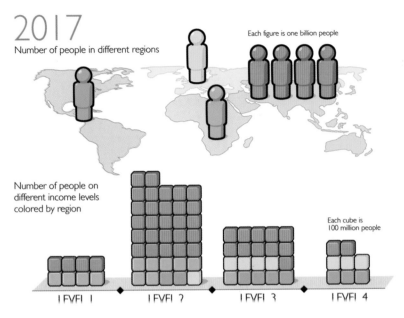

Number of people on
different income levels
colored by region

Each cube is
100 million people

LEVEL 1   LEVEL 2   LEVEL 3   LEVEL 4

Assuming that current trends continue, this is what the world might look like in 2040.

2040

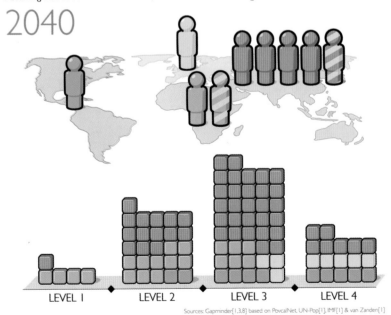

LEVEL 1   LEVEL 2   LEVEL 3   LEVEL 4

Sources: Gapminder[1,3,8] based on PovcalNet, UN-Pop[1], IMF[1] & van Zanden[1]

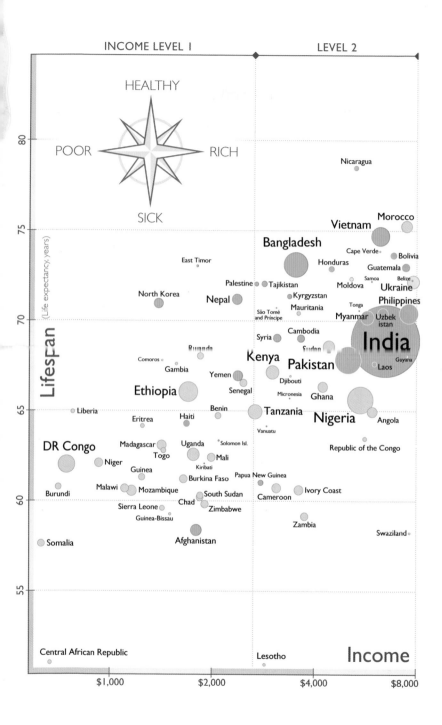

FACTFULNESS

FACTFULNESS

TEN REASONS
WE'RE WRONG
ABOUT THE WORLD—
AND WHY THINGS
ARE BETTER THAN
YOU THINK

Hans Rosling

with Ola Rosling and
Anna Rosling Rönnlund

FLATIRON
BOOKS
NEW YORK

FACTFULNESS. Copyright © 2018 by Factfulness AB. All rights reserved. Printed in the United States of America. For information, address Flatiron Books, 120 Broadway, New York, NY 10271.

www.flatironbooks.com

Designed by Steven Seighman

Illustrations and charts are based on free material from the Gapminder Foundation, designed by Ola Rosling and Anna Rosling Rönnlund.

The Library of Congress has cataloged the hardcover edition as follows:

Names: Rosling, Hans, author.
Title: Factfulness : ten reasons we're wrong about the world—
   and why things are better than you think / Hans Rosling,
   with Ola Rosling and Anna Rosling Rönnlund.
Description: First edition. | New York : Flatiron Books, 2018. |
   Includes bibliographical references and index.
Identifiers: LCCN 2018000167 | ISBN 9781250107817 (hardcover) | ISBN
   9781250231987 (international, sold outside the U.S., subject to rights
   availability) | ISBN 9781250123817 (ebook)
Subjects: LCSH: Stress management. | Reality.
Classification: LCC RA785 .R673 2018 | DDC 155.9'042—dc23
LC record available at https://lccn.loc.gov/2018000167

ISBN 978-1-250-12382-4 (trade paperback)

Our books may be purchased in bulk for promotional, educational, or business use. Please contact your local bookseller or the Macmillan Corporate and Premium Sales Department at 1-800-221-7945, extension 5442, or by email at MacmillanSpecialMarkets@macmillan.com.

First Flatiron Books Paperback Edition: April 2020

10 9 8 7 6 5 4 3 2

To the brave barefoot woman,
whose name I don't know but whose rational arguments
saved me from being sliced
by a mob of angry men with machetes

CONTENTS

AUTHOR'S NOTE

Factfulness is written in my voice, as if by me alone, and tells many stories from my life. But please don't be misled. Just like the TED talks and lectures I have been giving all over the world for the past ten years, this book is the work of three people, not one.

I am usually the front man. I stand onstage and deliver the lectures. I receive the applause. But everything you hear in my lectures, and everything you read in this book, is the output of eighteen years of intense collaboration between me, my son Ola Rosling, and my daughter-in-law Anna Rosling Rönnlund.

In 2005 we founded the Gapminder Foundation, with a mission to fight devastating ignorance with a fact-based worldview. I brought energy, curiosity, and a lifetime of experience as a doctor, a researcher, and a lecturer in global health. Ola and Anna were responsible for the data analysis, inventive visual explanations, data stories, and simple presentation design. It was their idea to measure ignorance systematically, and they designed and programmed our beautiful animated bubble charts. Dollar Street, a way of using photographs as data to explain the world, was Anna's brainchild. While I was getting ever angrier about people's ignorance about the world, Ola and Anna instead took the analysis beyond anger and crystallized the humble and relaxing idea of Factfulness. Together we defined the practical thinking tools that we present in this book.

What you are about to read was not invented according to the "lone genius" stereotype. It is instead the result of constant discussion, argument, and collaboration between three people with different talents, knowledge, and perspectives. This unconventional, often infuriating, but deeply productive way of working has led to a way of presenting the world and how to think about it, that I never could have created on my own.

INTRODUCTION

Why I Love the Circus

I love the circus. I love to watch a juggler throwing screaming chain saws in the air, or a tightrope walker performing ten flips in a row. I love the spectacle and the sense of amazement and delight at witnessing the seemingly impossible.

When I was a child my dream was to become a circus artist. My parents' dream, though, was for me to get the good education they never had. So I ended up studying medicine.

One afternoon at medical school, in an otherwise dry lecture about the way the throat worked, our professor explained, "If something is stuck, the passage can be straightened by pushing the chin bone forward." To illustrate, he showed an X-ray of a sword swallower in action.

I had a flash of inspiration.

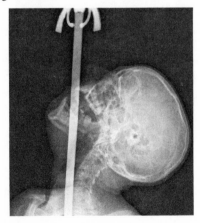

My dream was not over! A few weeks earlier, when studying reflexes, I had discovered that of all my classmates, I could push my fingers farthest down my throat without gagging. At the time, I had not been too proud: I didn't think it was an important skill. But now I understood its value, and instantly my childhood dream sprang back to life. I decided to become a sword swallower.

My initial attempts weren't encouraging. I didn't own a sword so used a fishing rod instead, but no matter how many times I stood in front of the bathroom mirror and tried, I'd get as far as an inch and it would get stuck. Eventually, for a second time, I gave up on my dream.

Three years later I was a trainee doctor on a real medical ward. One of my first patients was an old man with a persistent cough. I would always ask what my patients did for a living, in case it was relevant, and it turned out he used to swallow swords. Imagine my surprise when this patient turned out to be the very same sword swallower from the X-ray! And imagine this, when I told him all about my attempts with the fishing rod. "Young doctor," he said, "don't you know the throat is flat? You can only slide flat things down there. That is why we use a sword."

That night after work I found a soup ladle with a straight flat handle and immediately resumed my practice. Soon I could slide the handle all the way down my throat. I was excited, but being a soup ladle shaft swallower was not my dream. The next day, I put an ad in the local paper and soon I had acquired what I needed: a Swedish army bayonet from 1809. As I successfully slid it down my throat, I felt both deeply proud of my achievement and smug that I had found such a great way to recycle weapons.

Sword swallowing has always shown that the seemingly impossible can be possible, and inspired humans to think beyond the obvious. Occasionally I demonstrate this ancient Indian art at the end of one of my lectures on global development. I step up onto a table and rip off my professorial checked shirt to reveal a black vest top decorated with a gold sequined lightning bolt. I call for complete silence, and to the swirling beat of a snare drum I slowly slide the army bayonet down my throat. I stretch out my arms. The audience goes wild.

Test Yourself

This book is about the world, and how to understand it. So why start with the circus? And why would I end a lecture by showing off in a sparkly top? I'll soon explain. But first, I would like you to test your knowledge about the world. Please find a piece of paper and a pencil and answer the 13 fact questions below.

1. In all low-income countries across the world today, how many girls finish primary school?
☐ A: 20 percent
☐ B: 40 percent
☐ C: 60 percent

2. Where does the majority of the world population live?
☐ A: Low-income countries
☐ B: Middle-income countries
☐ C: High-income countries

3. In the last 20 years, the proportion of the world population living in extreme poverty has...
☐ A: almost doubled
☐ B: remained more or less the same
☐ C: almost halved

4. What is the life expectancy of the world today?
☐ A: 50 years
☐ B: 60 years
☐ C: 70 years

5. There are 2 billion children in the world today, aged 0 to 15 years old. How many children will there be in the year 2100, according to the United Nations?

□ A: 4 billion
□ B: 3 billion
□ C: 2 billion

6. The UN predicts that by 2100 the world population will have increased by another 4 billion people. What is the main reason?

□ A: There will be more children (age below 15)
□ B: There will be more adults (age 15 to 74)
□ C: There will be more very old people (age 75 and older)

7. How did the number of deaths per year from natural disasters change over the last hundred years?

□ A: More than doubled
□ B: Remained about the same
□ C: Decreased to less than half

8. There are roughly 7 billion people in the world today. Which map shows best where they live? (Each figure represents 1 billion people.)

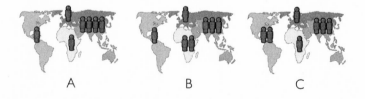

A   B   C

9. How many of the world's 1-year-old children today have been vaccinated against some disease?

□ A: 20 percent
□ B: 50 percent
□ C: 80 percent

10. Worldwide, 30-year-old men have spent 10 years in school, on average. How many years have women of the same age spent in school?

☐   A: 9 years
☐   B: 6 years
☐   C: 3 years

11. In 1996, tigers, giant pandas, and black rhinos were all listed as endangered. How many of these three species are more critically endangered today?

☐   A: Two of them
☐   B: One of them
☐   C: None of them

12. How many people in the world have some access to electricity?

☐   A: 20 percent
☐   B: 50 percent
☐   C: 80 percent

13. Global climate experts believe that, over the next 100 years, the average temperature will...

☐   A: get warmer
☐   B: remain the same
☐   C: get colder

Here are the correct answers:

1: C, 2: B, 3: C, 4: C, 5: C, 6: B, 7: C, 8: A, 9: C, 10: A, 11: C, 12: C, 13: A

Score one for each correct answer, and write your total score on your piece of paper.

Scientists, Chimpanzees, and You

How did you do? Did you get a lot wrong? Did you feel like you were doing a lot of guessing? If so, let me say two things to comfort you.

First, when you have finished this book, you will do much better. Not because I will have made you sit down and memorize a string of global statistics. (I am a global health professor, but I'm not crazy.) You'll do better because I will have shared with you a set of simple thinking tools. These will help you get the big picture right, and improve your sense of how the world works, without you having to learn all the details.

And second: if you did badly on this test, you are in very good company.

Over the past decades I have posed hundreds of fact questions like these, about poverty and wealth, population growth, births, deaths, education, health, gender, violence, energy, and the environment—basic global patterns and trends—to thousands of people across the world. The tests are not complicated and there are no trick questions. I am careful only to use facts that are well documented and not disputed. Yet most people do extremely badly.

Question three, for example, is about the trend in extreme poverty. Over the past twenty years, the proportion of the global population living in extreme poverty has halved. This is absolutely revolutionary. I consider it to be the most important change that has happened in the world in my lifetime. It is also a pretty basic fact to know about life on Earth. But people do not know it. On average only 7 percent—less than one in ten!—get it right.

FACT QUESTION 3 RESULTS: percentage who answered correctly.
In the last 20 years, the proportion of the world population living in extreme poverty has ... ?
(Correct answer: almost halved.)

Country	Percentage
Sweden	25%
Norway	25%
Finland	17%
Japan	10%
UK	9%
Canada	9%
Australia	6%
Germany	6%
US	5%
Belgium	5%
S. Korea	4%
France	4%
Spain	3%
Hungary	2%

0% — 100%

Sources: Ipsos MORI[1] & Novus[1]

(Yes, I have been talking a lot about the decline of global poverty in the Swedish media.)

The Democrats and Republicans in the United States often claim that their opponents don't know the facts. If they measured their own knowledge instead of pointing at each other, maybe everyone could become more humble. When we polled in the United States, only 5 percent picked the right answer. The other 95 percent, regardless of their voting preference, believed either that the extreme poverty rate had not changed over the last 20 years, or, worse, that it had actually doubled—which is literally the opposite of what has actually happened.

Let's take another example: question nine, about vaccination. Almost all children are vaccinated in the world today. This is amazing. It means that almost all human beings alive today have some access to basic modern health care. But most people do not know this. On average just 13 percent of people get the answer right.

FACT QUESTION 9 RESULTS: percentage who answered correctly.
How many of the world's one-year-old children today have been vaccinated against some disease?
(Correct answer: 80%.)

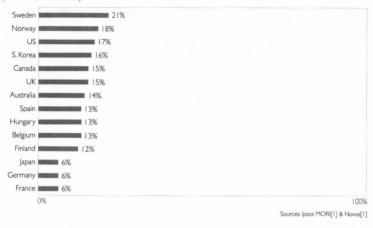

Sources: Ipsos MORI[1] & Novus[1]

Eighty-six percent of people get the final question about climate change right. In all the rich countries where we have tested public knowledge in online polls, most people know that climate experts are predicting warmer weather. In just a few decades, scientific findings have gone from the lab to the public. That is a big public-awareness success story.

Climate change apart though, it is the same story of massive ignorance (by which I do not mean stupidity, or anything intentional, but simply the lack of correct knowledge) for all twelve of the other questions. In 2017 we asked nearly 12,000 people in 14 countries to answer our questions. They scored on average just two correct answers out of the first 12. No one got full marks, and just one person (in Sweden) got 11 out of 12. A stunning 15 percent scored zero.

Perhaps you think that better-educated people would do better? Or people who are more interested in the issues? I certainly thought that once, but I was wrong. I have tested audiences from all around the world and from all walks of life: medical students, teachers, university lecturers, eminent scientists, investment bankers, executives in multinational companies, journalists, activists, and even senior political decision makers. These are highly educated people who take an interest in the world. But most of them—a stunning *majority* of them—get most of the answers wrong. Some of these groups even score *worse* than the general public; some of the most appalling results came from a group of Nobel laureates and medical researchers. It is not a question of intelligence. Everyone seems to get the world devastatingly wrong.

Not only devastatingly wrong, but *systematically* wrong. By which I mean that these test results are not random. They are worse than random: they are worse than the results I would get if the people answering my questions had no knowledge at all.

Imagine I decide to head down to the zoo to test out my questions on the chimpanzees. Imagine I take with me huge armfuls of bananas, each marked either A, B, or C, and throw them into the chimpanzee enclosure. Then I stand outside the enclosure, read out each question in a loud, clear voice, and note down, as each chimpanzee's "answer," the letter on the banana she next chooses to eat.

If I did this (and I wouldn't ever actually do this, but just imagine), the chimps, by picking randomly, would do consistently better than the well-educated but deluded human beings who take my tests. Through pure luck, the troop of chimps would score 33 percent on each three-answer question, or four out of the first 12 on the whole test. Remember that the humans I have tested get on average just two out of 12 on the same test.

What's more, the chimps' errors would be equally shared between the two wrong answers, whereas the human errors all tend to be in one direction. Every group of people I ask thinks the world is more frightening, more violent, and more hopeless—in short, more dramatic—than it really is.

Why Don't We Beat the Chimpanzees?

How can so many people be so wrong about so much? How is it even possible that the majority of people score worse than chimpanzees? *Worse than random!*

When I got my first little glimpse of this massive ignorance, back in the mid-1990s, I was pleased. I had just started teaching a course in global health at Karolinska Institutet in Sweden and I was a little nervous. These students were incredibly smart; maybe they would already know everything I had to teach them? What a relief when I discovered that my students knew less about the world than chimpanzees.

But the more I tested people, the more ignorance I found, not only among my students but everywhere. I found it frustrating and worrying that people were so wrong about the world. When you use the GPS in your car, it is important that it is using the right information. You wouldn't trust it if it seemed to be navigating you through a different city than the one you were in, because you would know that you would end up in the wrong place. So how could policy makers and politicians solve global problems if they were operating on the wrong facts? How could business people make sensible decisions for their organizations if their worldview were upside down? And how could each person going about their life know which issues they should be stressed and worried about?

I decided to start doing more than just testing knowledge and exposing ignorance. I decided to try to understand why. Why was this ignorance about the world so widespread and so persistent? We are all wrong sometimes—even me, I will readily admit that—but how could so many people be wrong about so much? Why were so many people scoring worse than the chimps?

Working late one night at the university I had a eureka moment. I realized the problem couldn't simply be that people lacked the knowledge, because that would give randomly incorrect answers—chimpanzee answers—rather than worse-than-random, worse-than-chimpanzee, systematically wrong answers. Only actively wrong "knowledge" can make us score so badly.

Aha! I had it! What I was dealing with here—or so I thought, for many years—was an upgrade problem: my global health students, and all the other people who took my tests over the years, did have knowledge, but it was outdated, often several decades old. People had a worldview dated to the time when their teachers had left school.

So, to eradicate ignorance, or so I concluded, I needed to upgrade people's knowledge. And to do that, I needed to develop better teaching materials setting out the data more clearly. After I told Anna and Ola about my struggles over a family dinner, both of them got involved and started to develop animated graphs. I traveled the world with these elegant teaching tools. They took me to TED talks in Monterey, Berlin, and Cannes, to the boardrooms of multinational corporations like Coca-Cola and IKEA, to global banks and hedge funds, to the US State Department. I was excited to use our animated charts to show everyone how the world had changed. I had great fun telling everyone that they were emperors with no clothes, that they knew nothing about the world. We wanted to install the worldview upgrade in everyone.

But gradually, gradually, we came to realize that there was something more going on. The ignorance we kept on finding was not just an upgrade problem. It couldn't be fixed simply by providing clearer data animations or better teaching tools. Because even people who loved my lectures, I sadly realized, weren't really hearing them. They might indeed be inspired, momentarily, but after the lecture, they were still stuck in their old negative worldview. The new ideas just wouldn't take. Even straight after my presentations, I would hear people expressing beliefs about poverty or population growth that I had just proven wrong with the facts. I almost gave up.

Why was the dramatic worldview so persistent? Could the media be to blame? Of course I thought about that. But it wasn't the answer. Sure, the media plays a role, and I discuss that later, but we must not make them into a pantomime villain. We cannot just shout "boo, *hiss*" at the media.

I had a defining moment in January 2015, at the World Economic Forum in the small and fashionable Swiss town of Davos. One thousand of the world's most powerful and influential political and business leaders, entrepreneurs, researchers, activists, journalists, and even many high-ranking UN officials had queued for seats at the forum's main session on socioeconomic and sustainable development, featuring me, and Bill and Melinda Gates. Scanning the room as I stepped onto the stage, I noticed several heads of state and a former secretary-general of the UN. I saw heads of UN organizations, leaders of major multinational companies, and journalists I recognized from TV.

I was about to ask the audience three fact questions—about poverty, population growth, and vaccination rates—and I was quite nervous. If my audience *did* know the answers to my questions, then none of the rest of my slides, revealing with a flourish how wrong they were, and what they should have answered, would work.

I shouldn't have worried. This top international audience who would spend the next few days explaining the world to each other did indeed know more than the general public about poverty. A stunning 61 percent of them got it right. But on the other two questions, about future population growth and the availability of basic primary health care, they still did worse than the chimps. Here were people who had access to all the latest data and to advisers who could continuously update them. Their ignorance could not possibly be down to an outdated worldview. Yet even they were getting the basic facts about the world wrong.

After Davos, things crystallized.

Our Dramatic Instincts and the Overdramatic Worldview

So here is this book. It shares with you the conclusions I finally reached—based on years of trying to teach a fact-based worldview, and listening to how people misinterpret the facts even when they are right there in front of them—about why so many people, from members of the public to very smart, highly educated experts, score worse than chimpanzees on fact questions about the world. (And I will also tell you what you can do about it.) In short:

Think about the world. War, violence, natural disasters, man-made disasters, corruption. Things are bad, and it feels like they are getting worse, right? The rich are getting richer and the poor are getting poorer; and the number of poor just keeps increasing; and we will soon run out of resources unless we do something drastic. At least that's the picture that most Westerners see in the media and carry around in their heads. I call it the overdramatic worldview. It's stressful and misleading.

In fact, the vast majority of the world's population lives somewhere in the middle of the income scale. Perhaps they are not what we think of as middle class, but they are not living in extreme poverty. Their girls go to school, their children get vaccinated, they live in two-child families, and they want to go abroad on holiday, not as refugees. Step-by-step, year-by-year, the world is improving. Not on every single measure every single year, but as a rule. Though the world faces huge challenges, we have made tremendous progress. This is the fact-based worldview.

It is the overdramatic worldview that draws people to the most dramatic and negative answers to my fact questions. People constantly and intuitively refer to their worldview when thinking, guessing, or learning about the world. So if your worldview is wrong, then you will systematically make wrong guesses. But this overdramatic worldview is not caused simply by out-of-date knowledge, as I once thought. Even people with access to the latest information get the world wrong. And I am convinced it is not the fault of an evil-minded media, propaganda, fake news, or wrong facts.

My experience, over decades of lecturing, and testing, and listening to the ways people misinterpret the facts even when they are right in front of them, finally brought me to see that the overdramatic worldview is so difficult to shift because it comes from the very way our brains work.

Optical Illusions and Global Illusions

Look at the two horizontal lines below. Which line is longest?

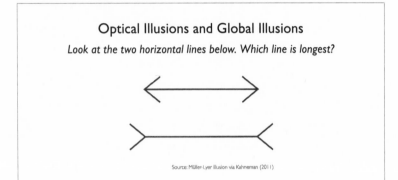

Source: Müller-Lyer illusion via Kahneman (2011)

You might have seen this before. The line on the bottom looks longer than the line on the top. You know it isn't, but even if you already know, even if you measure the lines yourself and confirm that they are the same, you keep seeing them as different lengths.

My glasses have a custom lens to correct for my personal sight problem. But when I look at this optical illusion, I still misinterpret what I see, just like everyone else. This is because illusions don't happen in our eyes, they happen in our brains. They are systematic misinterpretations, unrelated to individual sight problems. Knowing that most people are deluded means you don't need to be embarrassed. Instead you can be curious: how does the illusion work?

Similarly, you can look at the results from the public polls and skip being embarrassed. Instead be curious. How does this "global illusion" work? Why do so many people's brains systematically misinterpret the state of the world?

The human brain is a product of millions of years of evolution, and we are hard-wired with instincts that helped our ancestors to survive in small groups of hunters and gatherers. Our brains often jump to swift conclusions without much thinking, which used to help us to avoid immediate dangers. We are interested in gossip and dramatic stories, which used to be the only source of news and useful information. We crave sugar and fat, which used to be life-saving sources of energy when food was scarce. We have many instincts that used to be useful thousands of years ago, but we live in a very different world now.

Our cravings for sugar and fat make obesity one of the largest health problems in the world today. We have to teach our children, and ourselves, to stay away from sweets and chips. In the same way, our quick-thinking brains and cravings for drama—our dramatic instincts—are causing misconceptions and an overdramatic worldview.

Don't misunderstand me. We still need these dramatic instincts to give meaning to our world and get us through the day. If we sifted every input and analyzed every decision rationally, a normal life would be impossible. We should not cut out all sugar and fat, and we should not ask a surgeon to remove the parts of our brain that deal with emotions. But we need to learn to control our drama intake. Uncontrolled, our appetite for the dramatic goes too far, prevents us from seeing the world as it is, and leads us terribly astray.

Factfulness and the Fact-Based Worldview

This book is my very last battle in my lifelong mission to fight devastating global ignorance. It is my last attempt to make an impact on the world: to change people's ways of thinking, calm their irrational fears, and redirect their energies into constructive activities. In my previous battles I armed myself with huge data sets, eye-opening software, an energetic lecturing style, and a Swedish bayonet. It wasn't enough. But I hope that this book will be.

This is data as you have never known it: it is data as therapy. It is understanding as a source of mental peace. Because the world is not as dramatic as it seems.

Factfulness, like a healthy diet and regular exercise, can and should become part of your daily life. Start to practice it, and you will be able to replace your overdramatic worldview with a worldview based on facts. You will be able to get the world right without learning it by heart. You will make better decisions, stay alert to real dangers and possibilities, and avoid being constantly stressed about the wrong things.

I will teach you how to recognize overdramatic stories and give you some thinking tools to control your dramatic instincts. Then you will be able to shift your misconceptions, develop a fact-based worldview, and beat the chimps every time.

Back to the Circus

I occasionally swallow swords at the end of my lectures to demonstrate in a practical way that the seemingly impossible is possible. Before my circus act, I will have been testing my audience's factual knowledge about the world. I will have shown them that the world is completely different from what they thought. I will have proven to them that many of the changes they think will never happen have *already happened*. I will have been struggling to awaken their curiosity about what is possible, which is absolutely different from what they believe, and from what they see in the news every day.

I swallow the sword because I want the audience to realize how wrong their intuitions can be. I want them to realize that what I have shown them—both the sword swallowing and the material about the world that came before it—however much it conflicts with their preconceived ideas, however impossible it seems, is true.

I want people, when they realize they have been wrong about the world, to feel not embarrassment, but that childlike sense of wonder, inspiration, and curiosity that I remember from the circus, and that I still get every time I discover I have been wrong: "Wow, how is that even possible?"

This is a book about the world and how it really is. It is also a book about you, and why you (and almost everyone I have ever met) do not see the world as it really is. It is about what you can do about it, and how this will make you feel more positive, less stressed, and more hopeful as you walk out of the circus tent and back into the world.

So, if you are more interested in being right than in continuing to live in your bubble; if you are willing to change your worldview; if you are ready for critical thinking to replace instinctive reaction; and if you are feeling humble, curious, and ready to be amazed—then please read on.

CHAPTER ONE

THE GAP INSTINCT

*Capturing a monster in a classroom using only
a piece of paper*

Where It All Started

It was October 1995 and little did I know that after my class that evening, I was going to start my lifelong fight against global misconceptions.

"What is the child mortality rate in Saudi Arabia? Don't raise your hands. Just shout it out." I had handed out copies of tables 1 and 5 from UNICEF's yearbook. The handouts looked dull, but I was excited.

A choir of students shouted in unison: "THIRTY-FIVE."

"Yes. Thirty-five. Correct. This means that 35 children die before their fifth birthday out of every thousand live births. Give me the number now for Malaysia?"

"FOURTEEN," came the chorus.

As the numbers were thrown back at me, I scribbled them with a green pen onto a plastic film on the overhead projector.

"Fourteen," I repeated. "Fewer than Saudi Arabia!"

My dyslexia played a little trick on me and I wrote "Malaisya." The students laughed.

"Brazil?"

"FIFTY-FIVE."

"Tanzania?"

"ONE HUNDRED AND SEVENTY-ONE."

I put the pen down and said, "Do you know why I'm obsessed with the numbers for the child mortality rate? It's not *only* that I care about children. This measure takes the temperature of a whole society. Like a huge thermometer. Because children are very fragile. There are so many things that can kill them. When only 14 children die out of 1,000 in Malaysia, this means that the other 986 survive. Their parents and their society manage to protect them from all the dangers that could have killed them: germs, starvation, violence, and so on. So this number 14 tells us that most families in Malaysia have enough food, their sewage systems don't leak into their drinking water, they have good access to primary health care, and mothers can read and write. It doesn't just tell us about the health of children. It measures the quality of the whole society.

"It's not the numbers that are interesting. It's what they tell us about the lives behind the numbers," I continued. "Look how different these numbers are: 14, 35, 55, and 171. Life in these countries must be extremely different."

I picked up the pen. "Tell me now how life was in Saudi Arabia 35 years ago? How many children died in 1960? Look in the second column."

"TWO HUNDRED . . . and forty two."

The volume dropped as my students articulated the big number: 242.

"Yes. That's correct. Saudi Arabian society has made amazing progress, hasn't it? Child deaths per thousand dropped from 242 to

35 in just 33 years. That's way faster than Sweden. We took 77 years to achieve the same improvement.

"What about Malaysia? Fourteen today. What was it in 1960?"

"Ninety-three," came the mumbled response. The students had all started searching through their tables, puzzled and confused. A year earlier, I had given my students the same examples, but with no data tables to back them up, and they had simply refused to believe what I told them about the improvements across the world. Now, with all the evidence right in front of them, this year's students were instead rolling their eyes up and down the columns, to see if I had picked exceptional countries and tried to cheat them. They couldn't believe the picture they saw in the data. It didn't look anything like the picture of the world they had in their heads.

"Just so you know," I said, "you won't find any countries where child mortality has increased. Because the world in general is getting better. Let's have a short coffee break."

The Mega Misconception That "The World Is Divided in Two"

This chapter is about the first of our ten dramatic instincts, the gap instinct. I'm talking about that irresistible temptation we have to divide all kinds of things into two distinct and often conflicting groups, with an imagined gap—a huge chasm of injustice—in between. It is about how the gap instinct creates a picture in people's heads of a world split into two kinds of countries or two kinds of people: rich versus poor.

It's not easy to track down a misconception. That October evening in 1995 was the first time I got a proper look at the beast. It happened right after coffee, and the experience was so exciting that I haven't stopped hunting mega misconceptions ever since.

I call them mega misconceptions because they have such an enormous impact on how people misperceive the world. This first one is the worst. By dividing the world into two misleading boxes—poor and rich—it completely distorts all the global proportions in people's minds.

Hunting Down the First Mega Misconception

Starting up the lecture again, I explained that child mortality was highest in tribal societies in the rain forest, and among traditional farmers in the remote rural areas across the world. "The people you see in exotic documentaries on TV. Those parents struggle harder than anyone to make their families survive, and still they lose almost half of their children. Fortunately, fewer and fewer people have to live under such dreadful conditions."

A young student in the first row raised his hand. He tilted his head and said, "They can never live like us." All over the room other students nodded in support.

He probably thought I would be surprised. I was not at all. This was the same kind of "gap" statement I had heard many times before. I wasn't surprised, I was thrilled. This was what I had hoped for. Our dialogue went something like this:

ME: Sorry, who do you mean when you say "they"?
HIM: I mean people in other countries.
ME: All countries other than Sweden?
HIM: No. I mean . . . the non-Western countries. They can't live like us. It won't work.
ME: Aha! (As if now I understood.) You mean like Japan?
HIM: No, not Japan. They have a Western lifestyle.
ME: So what about Malaysia? They don't have a "Western lifestyle," right?

HIM: No. Malaysia is not Western. All countries that haven't adopted the Western lifestyle yet. They shouldn't. You know what I mean.

ME: No, I don't know what you mean. Please explain. You are talking about "the West" and "the rest." Right?

HIM: Yes. Exactly.

ME: Is Mexico . . . "West"?

He just looked at me.

I didn't mean to pick on him, but I kept going, excited to see where this would take us. Was Mexico "the West" and could Mexicans live like us? Or "the rest," and they couldn't? "I'm confused." I said. "You started with 'them and us' and then changed it to 'the West and the rest.' I'm very interested to understand what you mean. I have heard these labels used many times, but honestly I have never understood them."

Now a young woman in the third row came to his rescue. She took on my challenge, but in a way that completely surprised me. She pointed at the big paper in front of her and said, "Maybe we can define it like this: 'we in the West' have few children and few of the children die. While 'they in the rest' have many children and many of the children die." She was trying to resolve the conflict between his mind-set and my data set—in a pretty creative way, actually—by suggesting a definition for how to split the world. That made me so happy. Because she was absolutely wrong—as she would soon realize—and more to the point, she was wrong in a concrete way that I could test.

"Great. Fantastic. Fantastic." I grabbed my pen and leaped into action. "Let's see if we can put the countries in two groups based on how many children they have and how many children die."

The skeptical faces now became curious, trying to figure out what the heck had made me so happy.

I liked her definition because it was so clear. We could check it against the data. If you want to convince someone they are suffering from a misconception, it's very useful to be able to test their opinion against the data. So I did just that.

And I have been doing just that for the rest of my working life. The big gray photocopying machine that I had used to copy those original data tables was my first partner in my fight against misconceptions. By 1998, I had a new partner—a color printer that allowed me to share a colorful bubble graph of country data with my students. Then I acquired my first human partners, and things really picked up. Anna and Ola got so excited by these charts and my idea of capturing misconceptions that they joined my cause, and accidently created a revolutionary way to show hundreds of data trends as animated bubble charts. The bubble chart became our weapon of choice in our battle to dismantle the misconception that "the world is divided into two."

What's Wrong with This Picture?

My students talked about "them" and "us." Others talk about "the developing world" and "the developed world." You probably use these labels yourself. What's wrong with that? Journalists, politicians, activists, teachers, and researchers use them all the time.

When people say "developing" and "developed," what they are probably thinking is "poor countries" and "rich countries." I also often hear "West/rest," "north/south," and "low-income/high-income." Whatever. It doesn't really matter which terms people use to describe the world, as long as the words create relevant pictures in their heads and mean something with a basis in reality. But what pictures *are* in their heads when they use these two simple terms? And how do those pictures compare to reality?

Let's check against the data. The chart on the next page shows babies per woman and child survival rates for all countries.

Each bubble on the chart represents a country, with the size of the bubble showing the size of the country's population. The biggest bubbles are India and China. On the left of the chart are countries where women have many babies, and on the right are countries where women have few babies. The higher up a country is on the chart, the better the child survival rate in that country. This chart is exactly what my student in the third row suggested as a way of defining the two groups: "us and them," or "the West and the rest." Here I have labeled the two groups "developing and developed" countries.

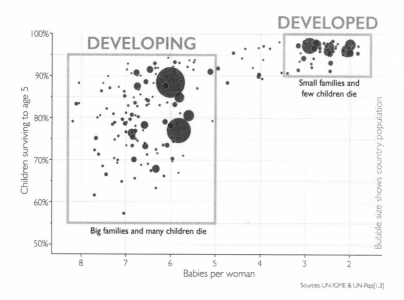

Look how nicely the world's countries fall into the two boxes: developing and developed. And between the two boxes there is a clear gap, containing just 15 small countries (including Cuba, Ireland, and Singapore) where just 2 percent of the world's population lives. In the box labeled "developing," there are 125 bubbles, including China and India. In all those countries, women have more than five children on average, and child deaths are common: fewer than 95 percent of children survive, meaning that more than 5 percent of children die before their fifth birthday. In the other box labeled "developed," there are 44 bubbles, including the United States and most of Europe. In all those countries the women have fewer than 3.5 children per woman and child survival is above 90 percent.

The world fits into two boxes. And these are exactly the two boxes that the student in the third row had imagined. This picture clearly shows a world divided into two groups, with a gap in the middle. How nice. What a simple world to understand! So what's the big deal? Why is it so wrong to label countries as "developed" and "developing"? Why did I give my student who referred to "us and them" such a hard time?

Because this picture shows the world in 1965! When I was a young man. That's the problem. Would you use a map from 1965 to navigate around your country? Would you be happy if your doctor was using cutting-edge research from 1965 to suggest your diagnosis and treatment? The picture below shows what the world looks like today.

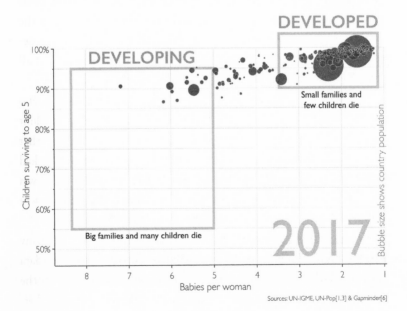

Sources: UN-IGME, UN-Pop[1,3] & Gapminder[6]

The world has completely changed. Today, families are small and child deaths are rare in the vast majority of countries, including the largest: China and India. Look at the bottom left-hand corner. The box is almost empty. The small box, with few children and high survival, that's where all countries are heading. And most countries are already there. Eighty-five percent of mankind are already inside the box that used to be named "developed world." The remaining 15 percent are mostly in between the two boxes. Only 13 countries, representing 6 percent of the world population, are still inside the "developing" box. But while the world has changed, the worldview has not, at least in the heads of the "Westerners." Most of us are stuck with a completely outdated idea about the rest of the world.

The complete world makeover I've just shown is not unique to family size and child survival rates. The change looks very similar for pretty much any aspect of human lives. Graphs showing levels of income, or tourism, or democracy, or access to education, health care, or electricity would all tell the same story: that the world used to be divided into two but isn't any longer. Today, most people are in the middle. There is no gap between the West and the rest, between developed and developing, between rich and poor. And we should all stop using the simple pairs of categories that suggest there is.

My students were dedicated, globally aware young people who wanted to make the world a better place. I was shocked by their blunt ignorance of the most basic facts about the world. I was shocked that they actually thought there were two groups, "us" and "them," shocked to hear them saying that "they" could not live like "us." How was it even possible that they were walking around with a 30-year-old worldview in their heads?

Pedaling home through the rain that evening in October 1995, my fingers numb, I felt fired up. My plan had worked. By bringing the data into the classroom I had been able to prove to my students that the world was not divided into two. I had finally managed to capture their misconception. Now I felt the urge to take the fight further. I realized I needed to make the data even clearer. That would help me to show more people, more convincingly, that their opinions were nothing more than unsubstantiated feelings. That would help me to shatter their illusions that they knew things that really they only felt.

Twenty years later I'm sitting in a fancy TV studio in Copenhagen in Denmark. The "divided" worldview is 20 years older, 20 years more outdated. We're live on air, and the journalist tilts his head and says to me, "We still see an enormous difference between the small, rich world, the old Western world mostly, and then the large part."

"But you're totally wrong," I reply.

Once more I explain that "poor developing countries" no longer exist as a distinct group. That there is no gap. Today, most people, 75 percent, live in middle-income countries. Not poor, not rich, but somewhere in the middle and starting to live a reasonable life. At one end of the scale there are still countries with a majority living in extreme and unacceptable poverty; at the other is the wealthy world (of North America and Europe and a few others like Japan, South Korea, and Singapore). But the vast majority are already in the middle.

"And what do you base that knowledge on?" continued the journalist in an obvious attempt to be provocative. And he succeeded. I couldn't help getting irritated and my agitation showed in my voice, and my words: "I use normal statistics that are compiled by the World Bank and the United Nations. This is not controversial. These facts are not up for discussion. I am right and you are wrong."

Capturing the Beast

Now that I have been fighting the misconception of a divided world for 20 years, I am no longer surprised when I encounter it. My students were not special. The Danish journalist was not special. The vast majority of the people I meet think like this. If you are skeptical about my claim that so many people get it wrong, that's good. You should always require evidence for claims like these. And here it is, in the form of a two-part misconception trap.

First, we had people disclose how they imagined life in so-called low-income countries, by asking questions like this one from the test you did in the introduction.

FACT QUESTION I

In all low-income countries across the world today, how many girls finish primary school?

☐   A: 20 percent
☐   B: 40 percent
☐   C: 60 percent

On average just 7 percent picked the correct answer, C: 60 percent of girls finish primary school in low-income countries. (Remember, 33 percent of the chimps at the zoo would have gotten this question right.) A majority of people "guessed" that it was just 20 percent. There are only a very few countries in the world—exceptional places like Afghanistan or South Sudan—where fewer than 20 percent of girls finish primary school, and at most 2 percent of the world's girls live in such countries.

When we asked similar questions about life expectancy, undernourishment, water quality, and vaccination rates—essentially asking what proportion of people in low-income countries had access to the basic first steps toward a modern life—we got the same kinds of results. Life expectancy in low-income countries is 62 years. Most people have enough to eat, most people have access to improved water, most children are vaccinated, and most girls finish primary school. Only tiny percentages—way less than the chimps' 33 percent—got

FACT QUESTION 1 RESULTS: percentage who answered correctly.
In all low-income countries across the world today, how many girls finish primary school?
(Correct answer: 60%.)

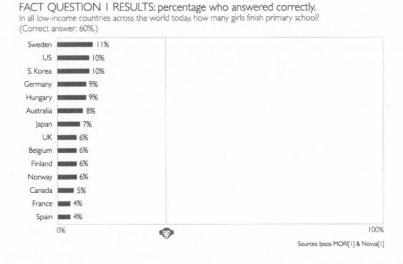

Sources: Ipsos MORI[1] & Novus[1]

these answers right, and large majorities picked the worst alternative
we offered, even when those numbers represented levels of misery
now being suffered only during terrible catastrophes in the very worst
places on Earth.

Now let's close the trap, and capture the misconception. We now
know that people believe that life in low-income countries is much
worse than it actually is. But how many people do they imagine live
such terrible lives? We asked people in Sweden and the United
States:

Of the world population, what percentage lives in low-income countries?

The majority suggested the answer was 50 percent or more. The
average guess was 59 percent.

The real figure is 9 percent. Only 9 percent of the world lives in
low-income countries. And remember, we just worked out that those
countries are not nearly as terrible as people think. They are really
bad in many ways, but they are not at or below the level of Afghanistan, Somalia, or Central African Republic, the worst places to live on
the planet.

To summarize: low-income countries are much more developed than most people think. And vastly fewer people live in them. The idea of a divided world with a majority stuck in misery and deprivation is an illusion. A complete misconception. Simply wrong.

Help! The Majority Is Missing

If the majority doesn't live in low-income countries, then where does it live? Surely not in high-income countries?

How do you like your bath water? Ice cold or steam hot? Of course, those are not the only alternatives. You can also have your water freezing, tepid, scalding, or anything in between. Many options, along a range.

FACT QUESTION 2

Where does the majority of the world population live?

☐ A: Low-income countries
☐ B: Middle-income countries
☐ C: High-income countries

FACT QUESTION 2 RESULTS: percentage who answered correctly.
Where does the majority of the world population live?
(Correct answer: middle-income countries.)

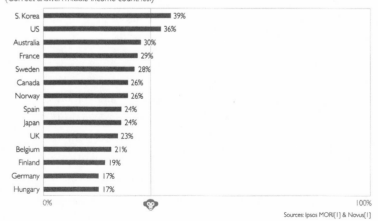

S. Korea	39%
US	36%
Australia	30%
France	29%
Sweden	28%
Canada	26%
Norway	26%
Spain	24%
Japan	24%
UK	23%
Belgium	21%
Finland	19%
Germany	17%
Hungary	17%

0%   100%

Sources: Ipsos MORI[1] & Novus[1]

The majority of people live neither in low-income countries nor in high-income countries, but in middle-income countries. This category doesn't exist in the divided mind-set, but in reality it definitely exists. It's where 75 percent of humanity lives, right there where the gap is supposed to be. Or, to put it another way, there is no gap.

Combining middle- and high-income countries, that makes 91 percent of humanity, most of whom have integrated into the global market and made great progress toward decent lives. This is a happy realization for humanitarians and a crucial realization for global businesses. There are 5 billion potential consumers out there, improving their lives in the middle, and wanting to consume shampoo, motorcycles, menstrual pads, and smartphones. You can easily miss them if you go around thinking they are "poor."

So What Should "We" Call "Them" Instead? The Four Levels

I am often quite rude about the term "developing countries" in my presentations.

Afterward, people ask me, "So what should we call them instead?" But listen carefully. It's the same misconception: we and them. What should "we" call "them" instead?

What we should do is stop dividing countries into two groups. It doesn't make sense anymore. It doesn't help us to understand the world in a practical way. It doesn't help businesses find opportunities, and it doesn't help aid money to find the poorest people.

But we need to do some kind of sorting to make sense of the world. We can't give up our old labels and replace them with . . . nothing. What should we do?

One reason the old labels are so popular is that they are so simple. But they are wrong! So, to replace them, I will now suggest an equally simple but more relevant and useful way of dividing up the world. Instead of dividing the world into two groups I will divide it into four income levels, as set out in the image below.

FOUR INCOME LEVELS

The world population in 2017. Billions of people on different income.

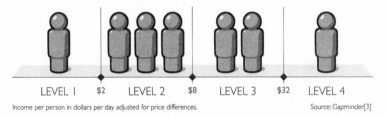

| LEVEL 1 | $2 | LEVEL 2 | $8 | LEVEL 3 | $32 | LEVEL 4 |

Income per person in dollars per day adjusted for price differences.   Source: Gapminder[3]

Each figure in the chart represents 1 billion people, and the seven figures show how the current world population is spread out across four income levels, expressed in terms of dollar income per day. You can see that most people are living on the two middle levels, where people have most of their basic human needs met.

Are you excited? You should be. Because the four income levels are the first, most important part of your new fact-based framework. They are one of the simple thinking tools I promised would help you to guess better about the world. Throughout the book you will see how the levels provide a simple way to understand all kinds of things, from terrorism to sex education. So I want to try to explain what life is like on each of these four levels.

Think of the four income levels as the levels of a computer game. Everyone wants to move from Level 1 to Level 2 and upward through the levels from there. Only, it's a very strange computer game, because Level 1 is the hardest. Let's play.

Water

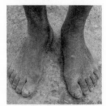

Transport

Cooking

Plate of food

LEVEL I $2

Source: Dollar Street

LEVEL 1. You start on Level 1 with $1 per day. Your five children have to spend hours walking barefoot with your single plastic bucket, back and forth, to fetch water from a dirty mud hole an hour's walk away. On their way home they gather firewood, and you prepare the same gray porridge that you've been eating at every meal, every day, for your whole life—except during the months when the meager soil yielded no crops and you went to bed hungry. One day your youngest daughter develops a nasty cough. Smoke from the indoor fire is weakening her lungs. You can't afford antibiotics, and one month later she is dead. This is extreme poverty. Yet you keep struggling on. If you are lucky and the yields are good, you can maybe sell some surplus crops and manage to earn more than $2 a day, which would move you to the next level. Good luck! (Roughly 1 billion people live like this today.)

Water

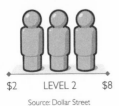

Transport

Cooking

Plate of food

LEVEL 2. You've made it. In fact, you've quadrupled your income and now you earn $4 a day. Three extra dollars every day. What are you going to do with all this money? Now you can buy food that you didn't grow yourself, and you can afford chickens, which means eggs. You save some money and buy sandals for your children, and a bike, and more plastic buckets. Now it takes you only half an hour to fetch water for the day. You buy a gas stove so your children can attend school instead of gathering wood. When there's power they do their homework under a bulb. But the electricity is too unstable for a freezer. You save up for mattresses so you don't have to sleep on the mud floor. Life is much better now, but still very uncertain. A single illness and you would have to sell most of your possessions to buy medicine. That would throw you back to Level 1 again. Another three dollars a day would be good, but to experience really drastic improvement you need to quadruple again. If you can land a job in the local garment industry you will be the first member of your family to bring home a salary. (Roughly 3 billion people live like this today.)

$2   LEVEL 2 $8

Source: Dollar Street

Water

Transport

Cooking

Plate of food

$8   LEVEL 3 $32

Source: Dollar Street

LEVEL 3. Wow! You did it! You work multiple jobs, 16 hours a day, seven days a week, and manage to quadruple your income again, to $16 a day. Your savings are impressive and you install a cold-water tap. No more fetching water. With a stable electric line the kids' homework improves and you can buy a fridge that lets you store food and serve different dishes each day. You save to buy a motorcycle, which means you can travel to a better-paying job at a factory in town. Unfortunately you crash on your way there one day and you have to use money you had saved for your children's education to pay the medical bills. You recover, and thanks to your savings you are not thrown back a level. Two of your children start high school. If they manage to finish, they will be able to get better-paying jobs than you have ever had. To celebrate, you take the whole family on its first-ever vacation, one afternoon to the beach, just for fun. (Roughly 2 billion people live like this today.)

Water

Transport

Cooking

Plate of food

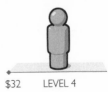

$32   LEVEL 4

Source: Dollar Street

LEVEL 4. You have more than $32 a day. You are a rich consumer and three more dollars a day makes very little difference to your everyday life. That's why you think three dollars, which can change the life of someone living in extreme poverty, is not a lot of money. You have more than twelve years of education and you have been on an airplane on vacation. You can eat out once a month and you can buy a car. Of course you have hot and cold water indoors.

But you know about this level already. Since you are reading this book, I'm pretty sure you live on Level 4. I don't have to describe it for you to understand. The difficulty, when you have always known this high level of income, is to understand the huge differences between the other three levels. People on Level 4 must struggle hard not to misunderstand the reality of the other 6 billion people in the world. (Roughly 1 billion people live like this today.)

I've described the progress up the levels as if one person managed to move through several levels. That is very unusual. Often it takes several generations for a family to move from Level 1 to Level 4. I hope though that you now have a clear picture of the kinds of lives people live on different levels; a sense that it is possible to move through the levels, both for individuals and for countries; and above all the understanding that there are not just two kinds of lives.

Human history started with everyone on Level 1. For more than 100,000 years nobody made it up the levels and most children didn't survive to become parents. Just 200 years ago, 85 percent of the world population was still on Level 1, in extreme poverty.

Today the vast majority of people are spread out in the middle, across Levels 2 and 3, with the same range of standards of living as people had in Western Europe and North America in the 1950s. And this has been the case for many years.

The Gap Instinct

The gap instinct is very strong. The first time I lectured to the staff of the World Bank was in 1999. I told them the labels "developing" and "developed" were no longer valid and I swallowed my sword. It took the World Bank 17 years and 14 more of my lectures before it finally announced publicly that it was dropping the terms "developing" and "developed" and would from now on divide the world into four income groups. The UN and most other global organizations have still not made this change.

So why is the misconception of a gap between the rich and the poor so hard to change?

I think this is because human beings have a strong dramatic instinct toward binary thinking, a basic urge to divide things into two distinct groups, with nothing but an empty gap in between. We love to dichotomize. Good versus bad. Heroes versus villains. My country versus the rest. Dividing the world into two distinct sides is simple and intuitive, and also dramatic because it implies conflict, and we do it without thinking, all the time.

Journalists know this. They set up their narratives as conflicts between two opposing people, views, or groups. They prefer stories of extreme poverty and billionaires to stories about the vast majority of people slowly dragging themselves toward better lives. Journalists are storytellers. So are people who produce documentaries and movies. Documentaries pit the fragile individual against the big, evil corporation. Blockbuster movies usually feature good fighting evil.

The gap instinct makes us imagine division where there is just a smooth range, difference where there is convergence, and conflict where there is agreement. It is the first instinct on our list because it's so common and distorts the data so fundamentally. If you look at the news or click on a lobby group's website this evening, you will probably notice stories about conflict between two groups, or phrases like "the increasing gap."

How to Control the Gap Instinct

There are three common warning signs that someone might be telling you (or you might be telling yourself) an overdramatic gap story and triggering your gap instinct. Let's call them comparisons of averages, comparisons of extremes, and the view from up here.

Comparisons of Averages

All you averages out there, please do not take offense at what I am about to say. I love averages. They are a quick way to convey information, they often tell us something useful, and modern societies couldn't function without them. Nor could this book. There will be many averages in this book. But any simplification of information may also be misleading, and averages are no exception. Averages mislead by hiding a spread (a range of different numbers) in a single number.

When we compare two averages, we risk misleading ourselves even more by focusing on the gap between those two single numbers, and missing the overlapping spreads, the overlapping ranges of numbers, that make up each average. That is, we see gaps that are not really there.

Look at the two (unrelated) graphs here, for example:

The graph on the left shows the gap between the average math scores of men and women taking SAT tests in the United States, for every year since 1965. The graph on the right shows the gap between the average income of people living in Mexico and the United States. Look at the huge differences between the two lines in each graph. Men versus women. United States versus Mexico. These graphs seem to show that men are better at math than women, and that people living in the United States have a higher income than Mexicans. And in a sense this is true. It is what the numbers say. But in what sense? To what extent? Are all men better than all women? Are all US citizens richer than all Mexicans?

Let's get a better sense of the reality behind the numbers. First, let's change the scale on the vertical axis. Using the same numbers, we now get a very different impression. Now the "gap" seems almost gone.

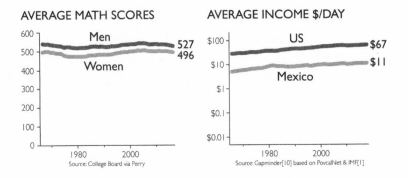

Now let's look at the same data in a third way. Instead of looking at the averages each year, let's look at the range of math scores, or incomes, in one particular year.

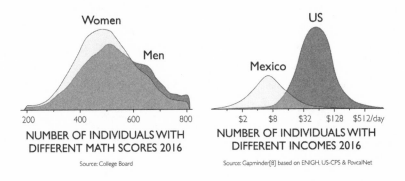

NUMBER OF INDIVIDUALS WITH
DIFFERENT MATH SCORES 2016

Source: College Board

NUMBER OF INDIVIDUALS WITH
DIFFERENT INCOMES 2016

Source: Gapminder[8] based on ENIGH, US-CPS & PovcalNet

Now we get a sense of all the individuals who were bundled into the average number. Look! There is an almost complete overlap between men and women's math scores. The majority of women have a male math twin: a man with the same math score as they do. When it comes to incomes in Mexico and the United States, the overlap is there but it is only partial. What is clear, though, looking at the data this way, is that the two groups of people—men and women, Mexicans and people living in the United States—are not separate at all. They overlap. There is no gap.

Of course, gap stories *can* reflect reality. In apartheid South Africa, black people and white people lived on different income levels and there was a true gap between them, with almost no overlap. The gap story of separate groups was absolutely relevant.

But apartheid was very unusual. Much more often, gap stories are a misleading overdramatization. In most cases there is no clear separation of two groups, even if it seems like that from the averages. We almost always get a more accurate picture by digging a little deeper and looking not just at the averages but at the spread: not just the group all bundled together, but the individuals. Then we often see that apparently distinct groups are in fact very much overlapping.

Comparisons of Extremes

We are naturally drawn to extreme examples, and they are easy to recall. For example, if we are thinking about global inequality we might think about the stories we have seen on the news about famine in South Sudan, on the one hand, and our own comfortable reality on the other. If we are asked to think about different kinds of government systems, we might quickly recall on the one hand corrupt, oppressive dictatorships and on the other hand countries like Sweden, with great welfare systems and benevolent bureaucrats dedicating their lives to safeguarding the rights of all citizens.

These stories of opposites are engaging and provocative and tempting—and very effective for triggering our gap instinct—but they rarely help understanding. There will always be the richest and the poorest, there will always be the worst regimes and the best. But the fact that extremes exist doesn't tell us much. The majority is usually to be found in the middle, and it tells a very different story.

Take Brazil, one of the world's most unequal countries. The richest 10 percent in Brazil earns 41 percent of the total income. Disturbing, right? It sounds too high. We quickly imagine an elite stealing resources from all the rest. The media support that impression with images of the very richest—often not the richest 10 percent but probably the richest 0.1 percent, the ultra-rich—and their boats, horses, and huge mansions.

Yes, the number is disturbingly high. At the same time, it hasn't been this low for many years.

SHARE OF TOTAL INCOME FOR BRAZIL'S RICHEST 10%

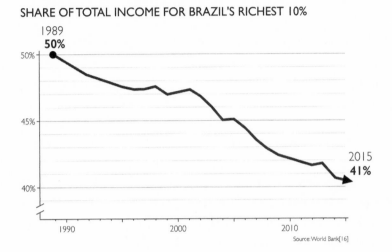

Source: World Bank[16]

Statistics are often used in dramatic ways for political purposes, but it's important that they also help us navigate reality. Let's now look at the incomes of the Brazilian population across the four levels.

NUMBER OF PEOPLE ON DIFFERENT INCOMES IN BRAZIL, 2016

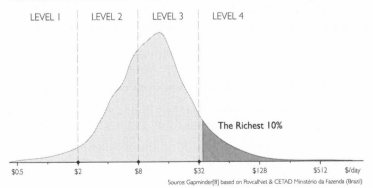

Source: Gapminder[8] based on PovcalNet & CETAD Ministério da Fazenda (Brazil)

Most people in Brazil have left extreme poverty. The big hump is on Level 3. That's where you get a motorbike and reading glasses, and save money in a bank to pay for high school and someday buy a washing machine. In reality, even in one of the world's most unequal countries, there is no gap. Most people are in the middle.

The View from Up Here

As I mentioned, if you are reading this book you probably live on Level 4. Even if you live in a middle-income country, meaning the average income is on Level 2 or 3—like Mexico, for example—you yourself probably live on Level 4 and your life is probably similar in important ways to the lives of the people living on Level 4 in San Francisco, Stockholm, Rio, Cape Town, and Beijing. The thing known as poverty in your country is different from "extreme poverty." It's "relative poverty." In the United States, for example, people are classified as below the poverty line even if they live on Level 3.

So the struggles people go through on Levels 1, 2, and 3 will most likely be completely unfamiliar to you. And they are not described in any helpful way in the mass media you consume.*

Your most important challenge in developing a fact-based worldview is to realize that most of your firsthand experiences are from Level 4; and that your secondhand experiences are filtered through the mass media, which loves nonrepresentative extraordinary events and shuns normality.

When you live on Level 4, everyone on Levels 3, 2, and 1 can look equally poor, and the word *poor* can lose any specific meaning. Even a person on Level 4 can appear poor: maybe the paint on their walls is peeling, or maybe they are driving a used car. Anyone who has

---

* Of course, if you live on Level 4 and have relatives living on Levels 2 or 3, you probably know what their lives look like. If so, you can skip this section.

looked down from the top of a tall building knows that it is difficult to assess from there the differences in height of the buildings nearer the ground. They all look kind of small. In the same way, it is natural for people living on Level 4 to see the world as divided into just two categories: rich (at the top of the building, like you) and poor (down there, not like you). It is natural to look down and say "oh, they are all poor." It is natural to miss the distinctions between the people with cars, the people with motorbikes and bicycles, the people with sandals, and the people with no shoes at all.

I assure you, because I have met and talked with people who live on every level, that for the people living on the ground on Levels 1, 2, and 3, the distinctions are crucial. People living in extreme poverty on Level 1 know very well how much better life would be if they could move from $1 a day to $4 a day, not to mention $16 a day. People who have to walk everywhere on bare feet know how a bicycle would save them tons of time and effort and speed them to the market in town, and to better health and wealth.

The four-level framework, the replacement for the overdramatic "divided" worldview, is the first and most important part of the fact-based framework you will learn in this book. Now you have learned it. It isn't too difficult, is it? I will use the four levels throughout the rest of the book to explain all kinds of things, including elevators, drownings, sex, cookery, and rhinos. They will help you to see the world more clearly and get it right more often.

What do you need to hunt, capture, and replace misconceptions? Data. You have to show the data and describe the reality behind it. So thank you, UNICEF data tables, thank you, bubble graphs, and thank you, internet. But you also need something more. Misconceptions disappear only if there is some equally simple but more relevant way of thinking to replace them. That's what the four levels do.

Factfulness

Factfulness is . . . recognizing when a story talks about a gap, and remembering that this paints a picture of two separate groups, with a gap in between. The reality is often not polarized at all. Usually the majority is right there in the middle, where the gap is supposed to be.

To control the gap instinct, **look for the majority.**

- **Beware comparisons of averages.** If you could check the spreads you would probably find they overlap. There is probably no gap at all.
- **Beware comparisons of extremes.** In all groups, of countries or people, there are some at the top and some at the bottom. The difference is sometimes extremely unfair. But even then the majority is usually somewhere in between, right where the gap is supposed to be.
- **The view from up here.** Remember, looking down from above distorts the view. Everything else looks equally short, but it's not.

CHAPTER TWO

THE NEGATIVITY INSTINCT

How I was kind of born in Egypt, and what a baby in an incubator can teach us about the world

Which statement do you agree with most?
- ☐ A: The world is getting better.
- ☐ B: The world is getting worse.
- ☐ C: The world is getting neither better nor worse.

Getting Out of the Ditch

I remember being suddenly upside down. I remember the dark, the smell of urine, and being unable to breathe as my mouth and nostrils filled with mud. I remember struggling to turn myself upright but only sinking deeper into the sticky liquid. I remember my arms, stretched out behind me, desperately searching the grass for something to pull, then being suddenly hauled out by the ankles. My grandma putting me in the big sink on the kitchen floor and washing me gently, with the hot water meant for the dishes. The scent of the soap.

These are my earliest memories and were nearly my last. They are memories of my rescue, aged four, from the sewage ditch running in front of my grandma's house. It was filled to the brim with a mix of last night's rain and sewage slurry from the factory workers' township. Something in it had caught my attention, and stepping to the ditch's edge, I had slipped and fallen in headfirst. My parents were not around to keep an eye on me. My mother was in the hospital, ill with tuberculosis. My father worked ten hours a day.

During the week, I lived with my grandparents. On Saturdays my daddy put me on the rack of his bike and we drove in large circles and figures of eight just for fun on our way to the hospital. I would see Mommy standing on the balcony on the third floor coughing. Daddy would explain that if we went in we could get sick too. I would wave to her and she would wave back. I saw her talking to me, but her voice was too weak and her words were carried away by the wind. I remember that she always tried to smile.

The Mega Misconception That "The World Is Getting Worse"

This chapter is about the negativity instinct: our tendency to notice the bad more than the good. This instinct is behind the second mega misconception.

"Things are getting worse" is the statement about the world that I hear more than any other. And it is absolutely true that there are many bad things in this world.

The number of war fatalities has been falling since the Second World War, but with the Syrian war, the trend has reversed. Terrorism too is rising again. (We'll get back to that in chapter 4.)

Overfishing and the deterioration of the seas are truly worrisome. The lists of dead areas in the world's oceans and of endangered species are getting longer.

Ice is melting. Sea levels will continue to rise by probably three feet over the next 100 years. There's no doubt it's because of all the greenhouse gases humans have pumped into the atmosphere, which won't disperse for a long time, even if we stop adding more.

The collapse of the US housing market in 2008, which no regulators had predicted, was caused by widespread illusions of safety in abstract investments, which hardly anyone understood. The system remains as complex now as it was then and a similar crisis could happen again. Maybe tomorrow.

In order for this planet to have financial stability, peace, and protected natural resources, there's one thing we can't do without, and that's international collaboration, based on a shared and fact-based understanding of the world. The current lack of knowledge about the world is therefore the most concerning problem of all.

I hear so many negative things all the time. Maybe you think, "Hans, you must just meet all the gloomiest people." We decided to check.

People in 30 countries and territories were asked the question at the top of the chapter: *Do you think the world is getting better, getting worse, or staying about the same?* This is what they said.

WHAT IS HAPPENING TO THE WORLD?
Percentage who answered "getting worse".
Overall, do you think the world is getting better, staying the same, or getting worse?

Source: YouGov[1] & Ipsos MORI[1]. See: gapm.io/rbetter

I never trust data 100 percent, and you never should either. There is always some uncertainty. In this case, I'd say these numbers are roughly right, but you shouldn't jump to any conclusions based on small differences. (By the way, that is a good general principle with statistics: be careful jumping to any conclusions if the differences are smaller than say, roughly, 10 percent.) The big picture is still crystal clear though.

The majority of people think the world is getting worse. No wonder we all feel so stressed.

Statistics as Therapy

It is easy to be aware of all the bad things happening in the world. It's harder to know about the good things: billions of improvements that are never reported. Don't misunderstand me, I'm not talking about some trivial positive news to supposedly balance out the negative. I'm talking about fundamental improvements that are world-changing but are too slow, too fragmented, or too small one-by-one to ever qualify as news. I'm talking about the secret silent miracle of human progress.

The basic facts about the world's progress are so little known that I get invited to talk about them at conferences and corporate meetings all over the world. They sometimes call my lectures "inspirational," and many people say they also have a comforting effect. That was never my intention. But it's logical. What I show is mostly just official UN data. As long as people have a worldview that is so much more negative than reality, pure statistics can make them feel more positive. It is comforting, as well as inspiring, to learn that the world is much better than you think. A new kind of happy pill, completely free online!

Extreme Poverty

Let's start by looking at the trend for extreme poverty.

FACT QUESTION 3

In the last 20 years, the proportion of the world population living in extreme poverty has...

☐   A: almost doubled
☐   B: remained more or less the same
☐   C: almost halved

The correct answer is C: over the last 20 years, the proportion of people living in extreme poverty has almost halved. But in our online polls, in most countries, less than 10 percent knew this.

Remember the four income levels from chapter 1? In the year 1800, roughly 85 percent of humanity lived on Level 1, in extreme poverty. All over the world, people simply did not have enough food. Most people went to bed hungry several times a year. Across Britain and its colonies, children had to work to eat, and the average child in the United Kingdom started work at age ten. One-fifth the entire Swedish population, including many of my relatives, fled starvation to the United States, and only 20 percent of them ever returned. When the harvest failed and your relatives, friends, and neighbors starved to death, what did you do? You escaped. You migrated. If you could.

Level 1 is where all of humanity started. It's where the majority always lived, until 1966. Until then, extreme poverty was the rule, not the exception.

EXTREME POVERTY RATE FROM 1800 TO TODAY

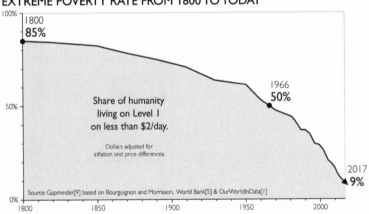

The curve you see above shows how the extreme poverty rate has been falling since 1800. And look at the last 20 years. Extreme poverty dropped faster than ever in world history.

In 1997, 42 percent of the population of both India and China were living in extreme poverty. By 2017, in India, that share had dropped to 12 percent: there were 270 million fewer people living in extreme poverty than there had been just 20 years earlier. In China, that share dropped to a stunning 0.7 percent over the same period, meaning another half a billion people over this crucial threshold. Meanwhile, Latin America took its proportion from 14 percent to 4 percent: another 35 million people. While all estimates of extreme poverty are very uncertain, when the change appears to be like this, then beyond all doubt something huge is happening.

How old were you 20 years ago? Close your eyes for a second and remember your younger self. How much has *your* world changed? A lot? A little? Well, this is how much *the* world has changed: just 20 years ago, 29 percent of the world population lived in extreme poverty. Now that number is 9 percent. Today almost everybody has escaped hell. The original source of all human suffering is about to be eradicated. We should plan a party! A big party! And when I say "we," I mean humanity!

Instead, we are gloomy. On our Level 4 TVs, we still see people in extreme poverty and it seems that nothing has changed. Billions of people have escaped misery and become consumers and producers for the world market, billions of people have managed to slide up from Level 1 to Levels 2 and 3, without the people on Level 4 noticing.

Life Expectancy

FACT QUESTION 4

What is the life expectancy of the world today?

- ☐ A: 50 years
- ☐ B: 60 years
- ☐ C: 70 years

Showing all the causes of deaths and suffering in one number is nearly impossible. But the average life expectancy gets very close. Every child death, every premature death from man-made or natural disasters, every mother dying in childbirth, and every elderly person's prolonged life is reflected in this measure.

Back in 1800, when Swedes starved to death and British children worked in coal mines, life expectancy was roughly 30 years everywhere in the world. That was what it had been throughout history. Among all babies who were ever born, roughly half died during their childhood. Most of the other half died between the ages of 50 and 70. So the average was around 30. It doesn't mean most people lived to be 30. It's just an average, and with averages we must always remember that there's a spread.

The average life expectancy across the world today is 70. Actually, it's better than that: it's 72. Here are the results of some polling.

FACT QUESTION 4 RESULTS: percentage who answered correctly.
What is the life expectancy of the world today? (Correct answer: 70 years.)
Country polls and selected lectures.

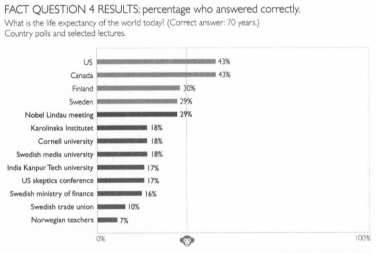

Sources: Ipsos MORI[1], Novus[1] & Gapminder[27]

This is one of those questions where the better educated you are, the more ignorant you seem to be. In most countries where we tested, members of the public just about beat the chimps. (The full country breakdown is in the appendix.) But in our more highly educated audiences, the most popular answer was 60 years. That would have been correct if we had asked the question in 1973 (the year when 200,000 people starved to death in Ethiopia). But we asked it in this decade, more than 40 years of progress later. People live on average ten years longer now. We humans have always struggled hard to make our families survive, and finally we are succeeding.

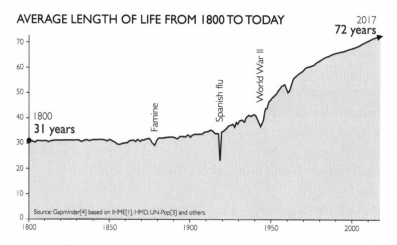

AVERAGE LENGTH OF LIFE FROM 1800 TO TODAY

When I show this amazing graph, people often ask, "What is the most recent dip there?," and they point at 1960. If you don't know already, this is a great opportunity for me to attack the misconception that the world is getting worse.

There's a dip in the global life expectancy curve in 1960 because 15 to 40 million people—nobody knows the exact number—starved to death that year in China, in what was probably the world's largest ever man-made famine.

The Chinese harvest in 1960 was smaller than planned because of a bad season combined with poor governmental advice about how to grow crops more effectively. The local governments didn't want to show bad results, so they took all the food and sent it to the central government. There was no food left. One year later the shocked inspectors were delivering eyewitness reports of cannibalism and dead bodies along roads. The government denied that its central planning had failed, and the catastrophe was kept secret by the Chinese government for 36 years. It wasn't described in English to the outside world until 1996. (Think about it. Could any government keep the death of 15 million people a global secret today?)

Even if the Chinese government had told the world about this tragedy, the UN World Food Programme—which today distributes food to wherever it is most needed in the world—couldn't have helped. It wasn't created until 1961.

The misconception that the world is getting worse is very difficult to maintain when we put the present in its historical context. We shouldn't diminish the tragedies of the droughts and famines happening right now. But knowledge of the tragedies of the past should help everyone realize how the world has become both much more transparent and much better at getting help to where it's needed.

I Was Born in Egypt

My home country of Sweden is today on Level 4 and one of the richest and healthiest countries in the world. (Saying that a country is on Level 4 means that the average person in that country is on Level 4. It doesn't mean that everyone in Sweden is on Level 4. Remember, averages disguise spreads.) But it hasn't always been so.

Now I'm going to show you my favorite graph. There's a color version of it on the inside front cover of this book. I call it the World Health Chart and it is like a world map for health and wealth. As with the bubble graph you saw in the previous chapter, each country is represented by a bubble, with the size of the bubble showing the size of the country's population. As before poorer countries are on the left and richer countries are on the right; healthier countries are higher up, and sicker countries are lower down.

Notice that there are not two groups. The world is not divided into two. There are countries on all levels, all the way from the sick and poor in the bottom left corner to the rich and healthy in the top right corner, where Sweden is. And most countries are in the middle.

Now this next bit is exciting.

The trail of little bubbles shows Sweden's health and wealth for every year since 1800. What tremendous progress! I have highlighted some countries that correspond, in 2017, to important years from Sweden's past.

SWEDEN'S HEALTH AND WEALTH FROM 1800 TO TODAY

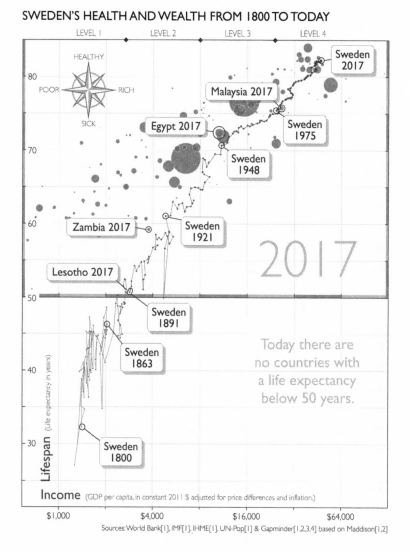

1948 was a very important year. The Second World War was over, Sweden topped the medals table at the Winter Olympics, and I was born. The Sweden I was born into in 1948 was where Egypt is on the health-wealth map today. That is to say, it was right in the middle of Level 3. Life conditions in 1950s Sweden were similar to those in Egypt or other countries on Level 3 today. There were still open sewage ditches and it wasn't uncommon for children to drown in bodies of water close to home. On Level 3, parents work hard, away from their children, and the government has not yet enforced regulations to protect water with fences.

Sweden kept improving during my lifetime. During the 1950s and 1960s it progressed all the way from Egypt today to Malaysia today. By 1975, the year Anna and Ola were born, Sweden, like Malaysia today, was just about to enter Level 4.

Let's go backward now. When my mother was born, in 1921, Sweden was like Zambia is now. That's Level 2.

My grandmother was the Lesothian member of our family. When she was born in 1891, Sweden was like Lesotho is today. That's the country with the shortest life expectancy in the world today, right on the border between Level 1 and 2, almost in extreme poverty. My grandmother hand-washed all the laundry for her family of nine all her adult life. But as she grew older, she witnessed the miracle of development as both she and Sweden reached Level 3. By the end of her life she had an indoor cold-water tap and a latrine bucket in the basement: luxury compared to her childhood, when there had been no running water. All four of my grandparents could spell and count, but none of them was literate enough to read for pleasure. They couldn't read children's books to me, nor could they write a letter. None of them had had more than four years of school. Sweden in my grandparents' generation had the same level of literacy that India, also on Level 2, has achieved today.

My great-grandmother was born in 1863, when Sweden's average income level was like today's Afghanistan, right on Level 1, with a majority of the population living in extreme poverty. Great-grandmother didn't forget to tell her daughter, my grandmother, how cold the mud floor used to be in the winter. But today people in Afghanistan and other countries on Level 1 live much longer lives than Swedes did back in 1863. This is because basic modernizations have reached most people and improved their lives drastically. They have plastic bags to store and transport food. They have plastic buckets to carry water and soap to kill germs. Most of their children are vaccinated. On average they live 30 years longer than Swedes did in 1800, when Sweden was on Level 1. That is how much life even on Level 1 has improved.

Your own country has been improving like crazy too. I can say this with confidence even though I don't know where you live, because every country in the world has improved its life expectancy over the last 200 years. In fact almost every country has improved by almost every measure.*

32 More Improvements

Is the world in your head still getting worse? Then get ready for a challenging data encounter. I have 32 more improvements to show you.

For each one, I could tell a similar story to those I have told about extreme poverty and life expectancy. For many of them I could show you that people are consistently more negative than the data says they should be. (And where I can't, it's because we haven't asked these questions yet.)

But I can't fit all these explanations into this book, so here are just the charts. Let's start with 16 terrible things that are on their way out, or have even already disappeared. And then, let's look at 16 wonderful things that have gotten better.

---

* You can track the progress of your country—or any country—using the freely available tool we use to create our bubble charts, found at www.gapminder.org/tools.

16 BAD THINGS DECREASING

LEGAL SLAVERY
Countries where forced labor is legal or practiced by the state (out of 195)

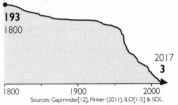

193
1800

2017
3

1800   1900   2000

Sources: Gapminder[12], Pinker (2011), ILO[1-5] & SDL.

OIL SPILLS
1,000 tonnes oil spilled from tanker ships

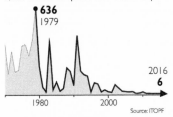

636
1979

2016
6

1980   2000

Source: ITOPF

EXPENSIVE SOLAR PANELS
Average price of PV modules ($/Wp)

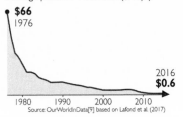

$66
1976

2016
$0.6

1980   1990   2000   2010

Source: OurWorldInData[9] based on Lafond et al. (2017)

HIV INFECTIONS
New HIV infections per million people

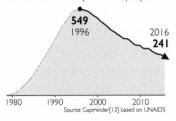

549
1996

2016
241

1980   1990   2000   2010

Source: Gapminder[13] based on UNAIDS

CHILDREN DYING
Percent dying before their fifth birthday

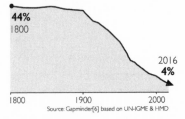

44%
1800

2016
4%

1800   1900   2000

Source: Gapminder[6] based on UN-IGME & HMD

BATTLE DEATHS
Battle deaths per 100,000 people

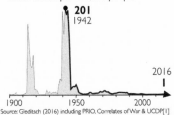

201
1942

2016
1

1900   1950   2000

Source: Gleditsch (2016) including PRIO, Correlates of War & UCDP[1]

DEATH PENALTY
Countries with death penalty (of 195)

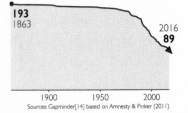

193
1863

2016
89

1900   1950   2000

Sources: Gapminder[14] based on Amnesty & Pinker (2011)

LEADED GASOLINE
Countries allowing lead in gasoline (of 195)

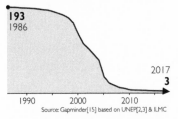

193
1986

2017
3

1990   2000   2010

Source: Gapminder[15] based on UNEP[2,3] & ILMC

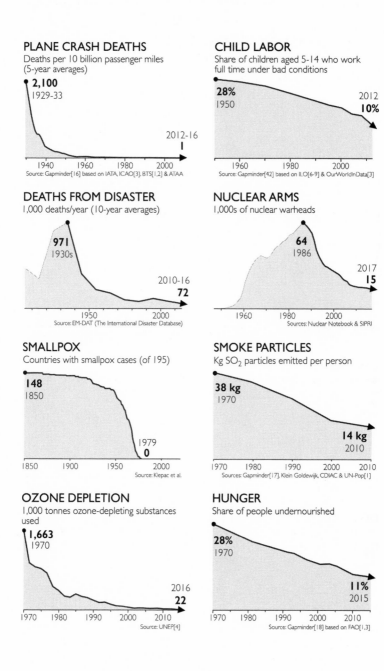

PLANE CRASH DEATHS
Deaths per 10 billion passenger miles
(5-year averages)

2,100
1929-33

2012-16
1

1940   1960   1980   2000
Source: Gapminder[16] based on IATA, ICAO[3], BTS[1,2] & ATAA

CHILD LABOR
Share of children aged 5-14 who work
full time under bad conditions

28%
1950

2012
10%

1960   1980   2000
Source: Gapminder[42] based on ILO[6-9] & OurWorldInData[3]

DEATHS FROM DISASTER
1,000 deaths/year (10-year averages)

971
1930s

2010-16
72

1950   2000
Source: EM-DAT (The International Disaster Database)

NUCLEAR ARMS
1,000s of nuclear warheads

64
1986

2017
15

1960   1980   2000
Sources: Nuclear Notebook & SiPRI

SMALLPOX
Countries with smallpox cases (of 195)

148
1850

1979
0

1850   1900   1950   2000
Source: Kiepac et al.

SMOKE PARTICLES
Kg SO$_2$ particles emitted per person

38 kg
1970

14 kg
2010

1970   1980   1990   2000   2010
Sources: Gapminder[17], Klein Goldewijk, CDIAC & UN-Pop[1]

OZONE DEPLETION
1,000 tonnes ozone-depleting substances
used

1,663
1970

2016
22

1970   1980   1990   2000   2010
Source: UNEP[4]

HUNGER
Share of people undernourished

28%
1970

11%
2015

1970   1980   1990   2000   2010
Source: Gapminder[18] based on FAO[1,3]

16 GOOD THINGS INCREASING

NEW MOVIES
Number of new feature films per year

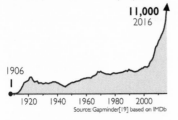

Source: Gapminder[19] based on IMDb

PROTECTED NATURE
Share of Earth's land surface protected as national parks and other reserves.

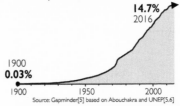

Source: Gapminder[5] based on Abouchakra and UNEP[5,6]

WOMEN'S RIGHT TO VOTE
Countries with equal rights for women and men to vote (out of 195)

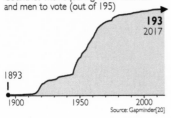

Source: Gapminder[20]

NEW MUSIC
New music recordings per year

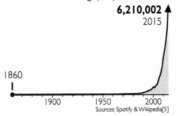

Sources: Spotify & Wikipedia[5]

SCIENCE
Scholarly articles published per year

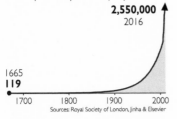

Sources: Royal Society of London, Jinha & Elsevier

HARVEST
Cereal yield (thousand kg per hectare)

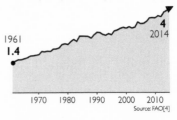

Source: FAO[4]

LITERACY
Share of adults (15+) with basic skills to read and write

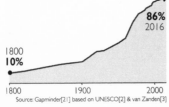

Source: Gapminder[21] based on UNESCO[2] & van Zanden[3]

DEMOCRACY
Share of humanity living in democracy

Source: OurWorldInData[4]

CHILD CANCER SURVIVAL

5-year survival of those diagnosed before age 20, with best treatment

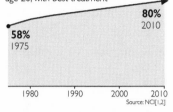

80%
2010

58%
1975

1980　1990　2000　2010

Source: NCI[1,2]

GIRLS IN SCHOOL

Share of girls of primary school age enrolled

90%
2015

65%
1970

1970　1980　1990　2000　2010

Source: UNESCO[3]

MONITORED SPECIES

Listed species with assessed threat-status

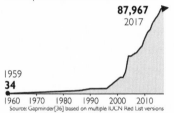

87,967
2017

1959
34

1960　1970　1980　1990　2000　2010

Source: Gapminder[36] based on multiple IUCN Red List versions

ELECTRICITY COVERAGE

Share of people with some access to electricity

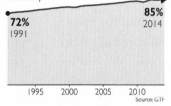

85%
2014

72%
1991

1995　2000　2005　2010

Source: GTF

MOBILE PHONES

Share of people with a cellphone

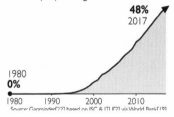

65%
2017

1980
0.0003%

1980　1990　2000　2010

Sources: GSMA & ITU[1]

WATER

Share of people with water from protected source

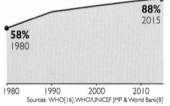

88%
2015

58%
1980

1980　1990　2000　2010

Sources: WHO[16], WHO/UNICEF JMP & World Bank[B]

INTERNET

Share of people using the Internet

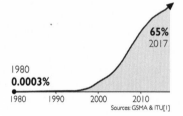

48%
2017

1980
0%

1980　1990　2000　2010

Source: Gapminder[72] based on ISC & ITU[2] via World Bank[19]

IMMUNIZATION

Share of 1-year-olds who got at least one vaccination

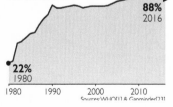

88%
2016

22%
1980

1980　1990　2000　2010

Sources: WHO[1] & Gapminder[23]

It is hard to see any of this global progress by looking out your window. It is taking place beyond the horizon. But there are some clues you can tune into, if you pay close attention. Listen carefully. Can you hear a child practicing the guitar or the piano? That child has not drowned, and is instead experiencing the joy and freedom of making music.

The goal of higher income is not just bigger piles of money. The goal of longer lives is not just extra time. The ultimate goal is to have the freedom to do what we want. Me, I love the circus, and playing computer games with my grandchildren, and zapping through TV channels. Culture and freedom, the goals of development, can be hard to measure, but guitars per capita is a good proxy. And boy, has that improved. With beautiful statistics like these, how can anyone say the world is getting worse?

GUITARS PER CAPITA
Playable guitars per million people

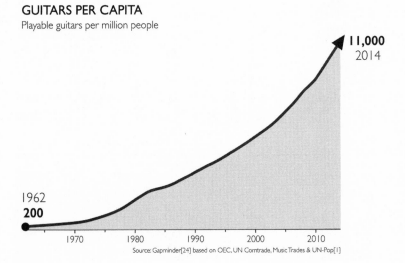

Source: Gapminder[24] based on OEC, UN Comtrade, Music Trades & UN-Pop[1]

The Negativity Instinct

In large part, it is because of our negativity instinct: our instinct to notice the bad more than the good. There are three things going on here: the misremembering of the past; selective reporting by journalists and activists; and the feeling that as long as things are bad it's heartless to say they are getting better.

Warning: Objects in Your Memories Were Worse Than They Appear

For centuries, older people have romanticized their youths and insisted that things ain't what they used to be. Well, that's true, but not in the way they mean it. Most things used to be worse, not better. But it is extremely easy for humans to forget how things really did "used to be."

In Western Europe and North America, only the very oldest, who lived through the Second World War or the Great Depression, have any personal recollection of the severe deprivation and hunger of just a few decades ago. Yet even in China and India, where extreme poverty was the reality for the vast majority just a couple of generations ago, it is now mostly forgotten by people who live in decent houses, have clean clothes, and ride mopeds.

The Swedish author and journalist Lasse Berg wrote an excellent report from rural India in the 1970s. When he returned 25 years later, he could see clearly how living conditions had improved. Pictures from his visit in the 1970s showed earthen floors, clay walls, half-naked children, and the eyes of villagers with low self-esteem and little knowledge of the outside world. They were a stark contrast to the concrete houses of the late 1990s, where well-dressed children played and self-confident and curious villagers watched TV. When Lasse showed the villagers the 1970s pictures they couldn't believe the photos were taken in their neighborhood. "No," they said. "This can't be here. You must be mistaken. We have never been that poor." Like most people, they were living in the moment, busy with new problems, like the children watching immoral soap operas or not having enough money to buy a motorbike.

Beyond living memory, for some reason we avoid reminding ourselves and our children about the miseries and brutalities of the past. The truth is to be found in ancient graveyards and burial sites, where archeologists have to get used to discovering that a large proportion of all the remains they dig up are those of children. Most will have been killed by starvation or disgusting diseases, but many child skeletons bear the marks of physical violence. Hunter-gatherer societies often had murder rates above 10 percent and children were not spared. In today's graveyards, child graves are rare.

Selective Reporting

We are subjected to never-ending cascades of negative news from across the world: wars, famines, natural disasters, political mistakes, corruption, budget cuts, diseases, mass layoffs, acts of terror. Journalists who reported flights that didn't crash or crops that didn't fail would quickly lose their jobs. Stories about gradual improvements rarely make the front page even when they occur on a dramatic scale and impact millions of people.

And thanks to increasing press freedom and improving technology, we hear more, about more disasters, than ever before. When Europeans slaughtered indigenous peoples across America a few centuries ago, it didn't make the news back in the old world. When central planning resulted in mass famine in rural China, millions starved to death while the youngsters in Europe waving communist red flags knew nothing about it. When in the past whole species or ecosystems were destroyed, no one realized or even cared. Alongside all the other improvements, our surveillance of suffering has improved tremendously. This improved reporting is itself a sign of human progress, but it creates the impression of the exact opposite.

At the same time, activists and lobbyists skillfully manage to make every dip in a trend appear to be the end of the world, even if the

MOST PEOPLE KEEP THINKING CRIME GOES UP

Gallup asked, "Is there more crime in the US than there was a year ago, or less?"

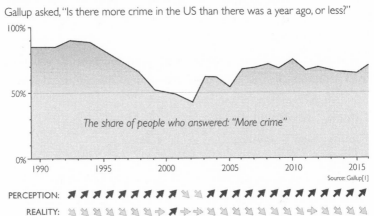

The share of people who answered: "More crime"

Source: Gallup[1]

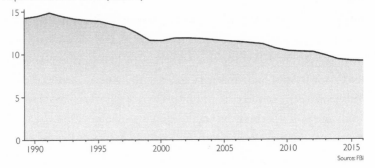

Reported crimes in US (millions)

Source: FBI

general trend is clearly improving, scaring us with alarmist exaggerations and prophecies. For example, in the United States, the crime rate has been on a downward trend since 1990. Just under 14.5 million crimes were reported in 1990. By 2016 that figure was well under 9.5 million. Each time something horrific or shocking happened, which was pretty much every year, a crisis was reported. The majority of people, the vast majority of the time, believe that crime is getting worse.

No wonder we get an illusion of constant deterioration. The news constantly alerts us to bad events in the present. The doom-laden feeling that this creates in us is then intensified by our inability to remember the past; our historical knowledge is rosy and pink and we fail to remember that, one year ago, or ten years ago, or 50 years ago, there was the same number of terrible events, probably more. This illusion of deterioration creates great stress for some people and makes other people lose hope. For no good reason.

Feeling, Not Thinking

There's something else going on as well. What are people really thinking when they say the world is getting worse? My guess is they are *not* thinking. They are feeling. If you still *feel* uncomfortable agreeing that the world is getting better, even after I have shown you all this beautiful data, my guess is that it's because you know that huge problems still remain. My guess is you *feel* that me saying that the world is getting better is like me telling you that everything is fine, or that you should look away from these problems and pretend they don't exist: and that feels ridiculous, and stressful.

I agree. Everything is not fine. We should still be very concerned. As long as there are plane crashes, preventable child deaths, endangered species, climate change deniers, male chauvinists, crazy dictators, toxic waste, journalists in prison, and girls not getting an education because of their gender, as long as any such terrible things exist, we cannot relax.

But it is just as ridiculous, and just as stressful, to look away from the progress that has been made. People often call me an optimist, because I show them the enormous progress they didn't know about. That makes me angry. I'm not an optimist. That makes me sound naïve. I'm a very serious "possibilist." That's something I made up. It means someone who neither hopes without reason, nor fears without reason, someone who constantly resists the overdramatic worldview. As a possibilist, I see all this progress, and it fills me with conviction and hope that further progress is possible. This is not optimistic. It is having a clear and reasonable idea about how things are. It is having a worldview that is constructive and useful.

When people wrongly believe that nothing is improving, they may conclude that nothing we have tried so far is working and lose confidence in measures that actually work. I meet many such people, who tell me they have lost all hope for humanity. Or, they may become radicals, supporting drastic measures that are counter-productive when, in fact, the methods we are already using to improve our world are working just fine.

Take, for example, girls' education. Educating girls has proven to be one of the world's best-ever ideas. When women are educated, all kinds of wonderful things happen in societies. The workforce becomes diversified and able to make better decisions and solve more problems. Educated mothers decide to have fewer children and more children survive. More energy and time is invested in each child's education. It's a virtuous cycle of change.

Poor parents who can't afford to send all their children to school have often prioritized the boys. But since 1970 there has been fantastic progress. Across religions, cultures, and continents, almost all parents can now afford to send all their children to school, and are sending their daughters as well as their sons. Now the girls have almost caught up: 90 percent of girls of primary school age attend school. For boys, the figure is 92 percent. There's almost no difference.

There are still gender differences when it comes to education on Level 1, especially when it comes to secondary and higher education, but that's no reason to deny the progress that has been made. I see no conflict between celebrating this progress and continuing to fight for more. I am a possibilist. And the progress we have made tells me it's possible to get all girls in school, and all boys too, and that we should work hard to make it happen. It won't happen by itself, and if we lose hope because of stupid misconceptions, it might not happen at all. The loss of hope is probably the most devastating consequence of the negativity instinct and the ignorance it causes.

How to Control the Negativity Instinct

How can we help our brains to realize that things are getting better when everything is screaming at us that things are getting worse?

Bad and Better

The solution is not to balance out all the negative news with more positive news. That would just risk creating a self-deceiving, comforting, misleading bias in the other direction. It would be as helpful as balancing too much sugar with too much salt. It would make things more exciting, but maybe even less healthy.

A solution that works for me is to persuade myself to keep two thoughts in my head at the same time.

It seems that when we hear someone say things are getting better, we think they are also saying "don't worry, relax" or even "look away."

But when I say things are getting better, I am not saying those things at all. I am certainly not advocating looking away from the terrible problems in the world. I am saying that things can be both bad and better.

Think of the world as a premature baby in an incubator. The baby's health status is extremely bad and her breathing, heart rate, and other important signs are tracked constantly so that changes for better or worse can quickly be seen. After a week, she is getting a lot better. On all the main measures, she is improving, but she still has to stay in the incubator because her health is still critical. Does it make sense to say that the infant's situation is improving? Yes. Absolutely. Does it make sense to say it is bad? Yes, absolutely. Does saying "things are improving" imply that everything is fine, and we should all relax and not worry? No, not at all. Is it helpful to have to choose between bad and improving? Definitely not. It's both. It's both bad and better. Better, and bad, at the same time.

That is how we must think about the current state of the world.

Expect Bad News

Something else that helps to control the negativity instinct is to constantly expect bad news.

Remember that the media and activists rely on drama to grab your attention. Remember that negative stories are more dramatic than neutral or positive ones. Remember how simple it is to construct a story of crisis from a temporary dip pulled out of its context of a long-term improvement. Remember that we live in a connected and transparent world where reporting about suffering is better than it has ever been before.

When you hear about something terrible, calm yourself by asking, If there had been an equally large positive improvement, would I have heard about that? Even if there had been hundreds of larger improvements, would I have heard? Would I ever hear about children who don't drown? Can I see a decrease in child drownings, or in deaths from tuberculosis, out my window, or on the news, or in a charity's publicity material? Keep in mind that the positive changes may be more common, but they don't find you. You need to find them. (And if you look in the statistics, they are everywhere.)

This reminder will give you the basic protection to allow you, and your children, to keep watching the news without being carried away into dystopia on a daily basis.

Don't Censor History

When we hang on to a rose-tinted version of history we deprive ourselves and our children of the truth. The evidence about the terrible past is scary, but it is a great resource. It can help us to appreciate what we have today and provide us with hope that future generations will, as previous generations did, get over the dips and continue the long-term trends toward peace, prosperity, and solutions to our global problems.

I Would Like to Thank . . . Society

Struggling for breath in that ditch full of pee 65 years ago in a working-class suburb in Sweden, little did I know that I would be the first in my family to go to university. Little did I know that I would become a global health professor and travel to Davos and tell the world's experts that they knew less about basic global trends than chimpanzees.

I didn't know any basic global trends myself back then, of course. I had to learn them. The only way anyone can know about different causes of death and how they are changing, for example, is to keep track of every death and its cause, write them down, and then add them up. That's extremely time-consuming. There's only one such data set in the whole world. It's named the Global Burden of Disease, and when I consulted it many years later it showed me that my near-death experience was not so special. It was a common type of accident for a child under five living on Level 3.

All I knew was that I was stuck. My grandmother came to the rescue and lifted me up. And then Swedish society lifted me further.

During my lifetime, Sweden moved from Level 3 to Level 4. A treatment against tuberculosis was invented and my mother got well. She read books to me that she borrowed from the public library. For free. I became the first in my family to get more than six years of education, and I went to university for free. I got a doctor's degree for free. Of course nothing is free: the taxpayers paid. And then, at the age of 30, when I had become a father of two and I discovered my first cancer, I was treated and cured by the world's best health-care system, for free. My survival and success in life have always depended on others. Thanks to my family, free education, and free health care, I made it all the way from that ditch to the World Economic Forum. I would never have made it on my own.

Today, now that Sweden is on Level 4, only three children in 1,000 die before the age of five, and only 1 percent of those deaths are drownings. Fences, day care, life-jacket campaigns, swimming lessons, and lifeguards at public pools all cost money. Child death from drowning is one of the many horrors that has nearly disappeared as the country has become richer. That is what I call progress. The same improvements are taking place across the world today. Most countries are currently improving faster than Sweden ever did. Much faster.

Factfulness

Factfulness is . . . recognizing when we get negative news, and re-membering that information about bad events is much more likely to reach us. When things are getting better we often don't hear about them. This gives us a systematically too-negative impression of the world around us, which is very stressful.

To control the negativity instinct, **expect bad news.**

- **Better and bad.** Practice distinguishing between a level (e.g., bad) and a direction of change (e.g., better). Convince yourself that things can be both better and bad.
- **Good news is not news.** Good news is almost never reported. So news is almost always bad. When you see bad news, ask whether equally positive news would have reached you.
- **Gradual improvement is not news.** When a trend is gradu-ally improving, with periodic dips, you are more likely to no-tice the dips than the overall improvement.
- **More news does not equal more suffering.** More bad news is sometimes due to better surveillance of suffering, not a worsening world.
- **Beware of rosy pasts.** People often glorify their early experi-ences, and nations often glorify their histories.

CHAPTER THREE

THE STRAIGHT LINE INSTINCT

How more survivors means fewer people,
how traffic accidents are like cavities, and
why my grandson is like the population of the world

The Most Frightening Graph I Ever Saw

Statistics can be terrifying. On September 23, 2014, I was sitting at my desk in the Gapminder office in Stockholm when I saw a line on a graph that gripped me with fear. I had been concerned about the Ebola outbreak in West Africa since August. Like others, I had seen the tragic images in the media of people dying in the streets of Monrovia, the capital of Liberia. But in my work, I often heard about sudden outbreaks of deadly diseases, and I had assumed it was like most others and would soon be contained. The graph in the World Health Organization research article shocked me into fear and then action.

The researchers had collected all the Ebola data since the start of the epidemic and used it to calculate the expected number of new cases per day up to the end of October. They showed, for the first time, that the number of cases was not just increasing along a straight line: 1, 2, 3, 4, 5. Instead, the number was doubling like this: 1, 2, 4, 8, 16. Each infected person was infecting, on average, two more people before dying. As a result, the number of new cases per day was doubling every three weeks. The graph showed how enormous the outbreak would soon become if each infected person kept infecting two more. Doubling is scary!

I had first learned about the effect of doubling at school. In the Indian legend, the Lord Krishna asks for one grain of rice on the first square of the chessboard, then two grains on the second square, four grains on the third square, then eight, and so on, doubling the number of grains each time. By the time he gets to the last of the 64 squares, he is owed 18,446,744,073,709,551,615 grains of rice: enough to cover the whole of India with a layer of rice 30 inches deep. Anything that keeps doubling grows much faster than we first assume. So I knew the situation in West Africa was about to become desperate. Liberia was at risk of a catastrophe worse than its recently ended civil war, and one that would almost inevitably spread to the entire world. Unlike malaria, Ebola could spread quickly in all climates and could travel on airplanes, across borders and oceans inside the bodies of unknowingly infected passengers. There was no effective treatment for it.

People were already dying in the streets now. Within only nine weeks (the time needed for three doublings) the situation would be eight times as desperate. Every three-week delay in dealing with the problem would mean twice as many people infected and twice as many resources needed. Ebola had to be stopped within weeks.

At Gapminder we immediately changed our priorities and started studying the data and producing information videos to explain the urgency of the situation. By October 20, I had canceled all my assignments for the next three months and was on a plane to Liberia, where I hoped my 20 years of studying epidemics in rural sub-Saharan Africa could be of some use. I remained in Liberia for three months, missing Christmas and New Year's with my family for the first time ever.

Like the rest of the world, I was too slow to understand the magnitude and urgency of the Ebola crisis. I had assumed that the increase in cases was a straight line when in fact the data clearly showed that it was a doubling line. Once I understood this, I acted. But I wish I had understood, and acted, sooner.

The Mega Misconception That "The World Population Is _Just_ Increasing and Increasing"

Nowadays, the word _sustainability_ is found in the title of almost every conference I get invited to. One of the most important numbers of the sustainability equation is the human population. There must be some kind of limit to how many people can live on this planet. Right? So when I started testing my audiences at these sustainability conferences, I just assumed that they would know the basic facts about global population growth. Seldom have I been so wrong.

We have now arrived at the third instinct—the straight line instinct—and the third and last mega misconception: the false idea that the world population is _just_ increasing. Please pay attention to the word _just_, which I've made italic and underlined for a purpose. This word is the misconception.

In fact, the world population _is_ increasing. Very fast. Roughly a billion people will be added over the next 13 years. That's true. That's not a misconception. But it's not _just_ increasing. The "just" implies that, if nothing is done, the population will just keep on growing. It implies that some drastic action is needed in order to stop the growth. That is the misconception, and I think it is based on the same instinct that stopped me and the world from acting sooner to stop Ebola. The instinct to assume that lines are straight.

I rarely get speechless, but it happened the first time I asked an audience the following question. It was at a teachers' conference in Norway (but I don't mean to be too hard on the Norwegians: it might just as well have been in Finland too). Many of these teachers were teaching global population trends as part of their social science classes. When I turned my head around and saw the results from the live poll on the screen behind me, I couldn't find words. I remember thinking that there must be something wrong with the polling devices.

FACT QUESTION 5

There are 2 billion children in the world today, aged 0 to 15 years old. How many children will there be in the year 2100, according to the UN?

NUMBER OF CHILDREN IN THE WORLD
Population aged 0 to 14 years

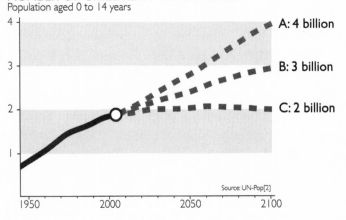

Source: UN-Pop[2]

Before asking the question, I had told the teachers, "One of these three lines shows the official UN forecast. The other two lines, I just made up."

Again, chimpanzees pick the correct line 33 percent of the time. The teachers in Norway? Only 9 percent. I was shocked. How could such an important group of people score worse than random? What were they teaching the children?

I kind of hoped the polling devices were broken. But they were not. We got the same terrible results in our public polls. In the United States, the United Kingdom, Sweden, Germany, France, and Australia, 85 percent of people picked the fake lines. (The full country breakdown is in the appendix.)

The experts at the World Economic Forum? They answered much better than the public. Almost as well as chimpanzees. Twenty-six percent got it right.

Thinking it over more calmly after the teachers' conference was over, I started to see the size of the knowledge problem. The number of future children is the most essential number for making global population forecasts. So it is central to the whole sustainability debate. If we get this number wrong, we are going to get a lot else wrong. Yet almost none of the highly educated and influential people we have measured have the slightest knowledge of what the population experts are all agreeing about. The numbers are freely available online, from the UN website, but free access to data doesn't turn into knowledge without effort. The UN line is alternative C: the flat line at the bottom. UN experts expect that in the year 2100 there will be 2 billion children, the same number as today. They don't expect the line to continue straight. They expect no further increase. I'll soon get back to this.

The Straight Line Instinct

This graph shows the world population since the year 8000 BC. That's when agriculture was invented.

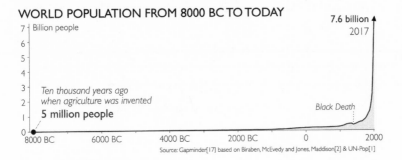

WORLD POPULATION FROM 8000 BC TO TODAY

7.6 billion
2017

Ten thousand years ago
when agriculture was invented
5 million people

Black Death

Source: Gapminder[17] based on Biraben, McEvedy and Jones, Maddison[2] & UN-Pop[1]

Back then, the total human population was roughly 5 million people, spread along coastlines and rivers all over the world. The total of humanity was smaller than the population of one of our big cities today: London, Bangkok, or Rio de Janeiro.

This number increased only slowly for almost 10,000 years, eventually reaching 1 billion in the year 1800. Then something happened. The next billion were added in only 130 years. And another 5 billion were added in under 100 years. Of course people get worried when they see such a steep increase, and they know the planet has limited resources. It sure *looks* like it's *just* increasing, and at a very high speed.

When looking at a stone flying toward you, you can often predict whether it is going to hit you. You need no numbers, no graphs, no spreadsheets. Your eyes and brain extend the trajectory and you move out of the stone's way. It's easy to imagine how this automatic visual forecasting skill helped our ancestors survive. And it still helps us survive: when driving a car, we constantly predict where other cars will be within the next few seconds.

But our straight line intuition is not always a reliable guide in modern life.

When looking at a line graph, for example, it's nearly impossible *not* to imagine a straight line that stretches beyond the end of the trend, into the future. On the population graph on the next page, I added the dashed line to clarify what I think people are instinctively imagining. Of course they get worried.

PERCEIVED WORLD POPULATION INTO THE FUTURE

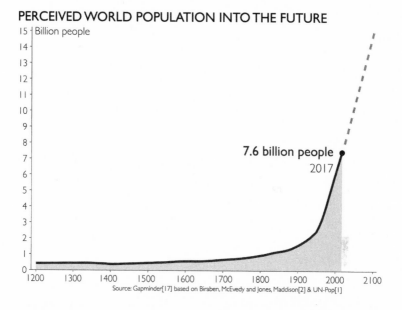

Source: Gapminder[17] based on Biraben, McEvedy and Jones, Maddison[2] & UN-Pop[1]

Let me now give you another example that I know you are more familiar with. My youngest grandchild, Mino, was 19.5 inches long when he was born. In his first six months he grew to 26.5 inches. An impressive growth of seven inches. Impressive, but also scary. Look at his growth chart. I have added the intuitive straight line into the future. It's terrifying, isn't it?

MINO'S HEIGHT INTO THE FUTURE

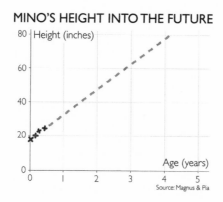

Source: Magnus & Pia

If Mino *just* continues growing, he will be 60 inches tall on his third birthday—a five-foot toddler. By his tenth birthday he will be 160 inches tall—over 13 feet. And then what? This can't *just* continue! Somebody must do something drastic! Mino's parents will have to remodel their house or find some medication!

The straight line intuition is obviously wrong in this case. Why is it obvious? Because we all have firsthand experience of a growing body. We know Mino's growth curve won't just continue. We've never met a person 160 inches tall. Assuming the trend will continue along a straight line is obviously ludicrous. But when we're less familiar with a topic, it's surprisingly difficult to imagine how stupid such an assumption may be.

The UN population experts have firsthand experience of calculating population sizes. It's their job. This is the line they expect:

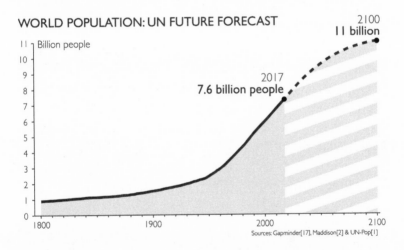

WORLD POPULATION: UN FUTURE FORECAST

Sources: Gapminder[17], Maddison[2] & UN-Pop[1]

The world population today is 7.6 billion people, and yes, it's growing fast. Still, the growth has already started to slow down, and the UN experts are pretty sure it will keep slowing down over the next few decades. They think the curve will flatten out at somewhere between 10 and 12 billion people by the end of the century.

The Shape of the Population Curve

To understand the shape of this population curve, we need to understand where the increase in population is coming from.

Why Is the Population Increasing?

FACT QUESTION 6

The UN predicts that by 2100 the world population will have increased by another 4 billion people. What is the main reason?

☐ A: There will be more children (age below 15)
☐ B: There will be more adults (age 15 to 74)
☐ C: There will be more very old people (age 75 and older)

This one, I'll give you the answer right away. The correct answer is B. The experts are convinced the population will keep growing, mainly because there will be more adults. Not more children and not more very old people. More adults. Here's the same population graph I just showed you, but now separating children and adults:

WORLD POPULATION: UN FUTURE FORECAST

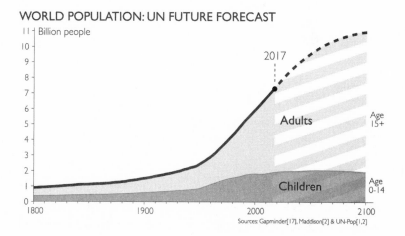

Sources: Gapminder[17], Maddison[2] & UN-Pop[1,2]

The number of children is not expected to increase, which we know already from this chapter's first fact question. Now look closely at the children line in this graph. Can you see when it gets flat? Can you see that it is already happening? The UN experts are not *predicting* that the number of children *will* stop increasing. They are *reporting* that it is already happening. The radical change that is needed to stop rapid population growth is that the number of children stops growing. And that is already happening. How could that be? That, everybody should know.

Attention, now! Because this next chart is the most dramatic in this book. It shows the incredible, truly world-changing drop in the number of babies per woman that has happened during my lifetime.

When I was born in 1948, women on average gave birth to five children each. After 1965 the number started dropping like it never had done before. Over the last 50 years it dropped all the way to the amazingly low world average of just below 2.5.

AVERAGE NUMBER OF BABIES PER WOMAN FROM 1800 TO TODAY

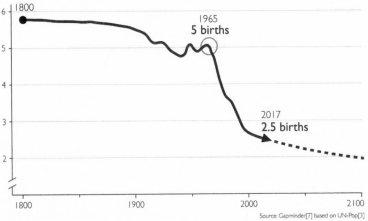

Source: Gapminder[7] based on UN-Pop[3]

This dramatic change happened in parallel with all those other improvements I described in the last chapter. As billions of people left extreme poverty, most of them decided to have fewer children. They no longer needed large families for child labor on the small family farm. And they no longer needed extra children as insurance against child mortality. Women and men got educated and started to want better-educated and better-fed children: and having fewer of them was the obvious solution. In practice, that goal was easier to realize thanks to the wonderful blessing of modern contraceptives, which let parents have fewer children without having less sex.

The dramatic drop in babies per woman is expected to continue, as long as more people keep escaping extreme poverty, and more women get educated, and as access to contraceptives and sexual education keeps increasing. Nothing drastic is needed. Just more of what we are already doing. The exact speed of the future drop is not possible to predict exactly. It depends on how fast these changes continue to happen. But in any case, the annual number of births in the world has already stopped increasing, which means that the period of fast population growth will soon be over. We are now arriving at "peak child."

But then, if the number of births has already stopped increasing, where are the 4 billion new adults going to come from? Spaceships?

Why Will the Population Stop Increasing?

The chart on the next page shows the population of the world divided into age groups, in 2015 and then every 15 years after that.

FUTURE WORLD POPULATION BY AGE GROUP

Each figure is 1 billion people.

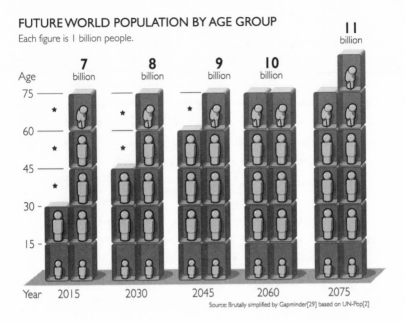

Source: Brutally simplified by Gapminder[29] based on UN-Pop[2]

On the left, the chart shows the ages of the 7 billion people alive in 2015: 2 billion were aged 0 to 15, 2 billion aged 15 to 30, and then there were 1 billion each in the 30 to 45, 45 to 60, and 60 to 75 age groups.

In 2030, there will be 2 billion new 0- to 15-year-olds. Everyone else will have grown older. The 0- to 15-year-olds of today will have become 15- to 30-year-olds. The 15- to 30-year-olds of today will have become 2 billion 30- to 45-year-olds. There are only 1 billion 30- to 45-year-olds today. So, without any increase in the number of children being born, and without people living for longer, there will be 1 billion more adults.

The 1 billion new adults come not from new children, but from children and young adults who have already been born.

For three generations, this pattern will repeat itself. In 2045, the 2 billion 30- to 45-year-olds will become 45- to 60-year-olds and we will have another 1 billion adults. In 2060, the 2 billion 45- to 60-year-olds will become 60- to 75-year-olds and we will have another 1 billion adults. But look what happens next. From 2060, each generation of 2 billion people will be replaced by another generation of 2 billion people. The fast growth stops.

The large increase in population is going to happen not because there are more children. And not, in the main, because old folks are living longer. In fact the UN experts do predict that by 2100, world life expectancy will have increased by roughly 11 years, adding 1 billion old people to the total and taking it to around 11 billion. The large increase in population will happen mainly because the children who already exist today are going to grow up and "fill up" the diagram with 3 billion more adults. This "fill-up effect" takes three generations, and then it is done.

That's actually all you need to know to understand the method that the UN experts use to not *just* draw a straight line into the future.

(This explanation is a brutal simplification. Many die before they reach 75, and many parents have their children after they reach 30. But even including these facts makes no difference to the big picture.)

In Balance with Nature

When a population is not growing over a long period of time, and the population curve is flat, this must mean that each generation of new parents is the same size as the previous one. For thousands of years up to 1800 the population curve was almost flat. Have you heard people say that humans used to live in balance with nature?

Well, yes, there was a balance. But let's avoid the rose-tinted glasses. Until 1800, women gave birth to six children on average. So the population should have increased with each generation. Instead, it stayed more or less stable. Remember the child skeletons in the grave-yards of the past? On average four out of six children died before becoming parents themselves, leaving just two surviving children to parent the next generation. There was a balance. It wasn't because humans *lived* in balance with nature. Humans *died* in balance with nature. It was utterly brutal and tragic.

Today, humanity is once again reaching a balance. The number of parents is no longer increasing. But this balance is dramatically different from the old balance. The new balance is nice: the typical parents have two children, and neither of them dies. For the first time in human history, we *live* in balance.

The population grew from 1.5 billion in 1900 to 6 billion in 2000 because humanity went through a transition from one balance to another during the twentieth century, a unique period of human history when two parents on average produced more than two children who survived to become parents themselves in the next generation.

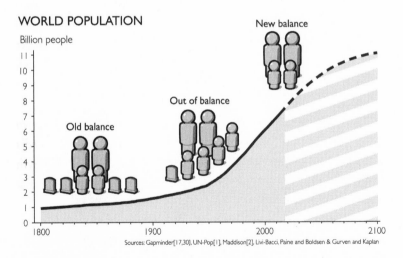

WORLD POPULATION

Billion people

New balance

Out of balance

Old balance

Sources: Gapminder[17,30], UN-Pop[1], Maddison[2], Livi-Bacci, Paine and Boldsen & Gurven and Kaplan

That period of imbalance is the reason why today the two youngest generations are larger than the others. That period of imbalance is the reason behind the fill-up. But the new balance is already achieved: the annual number of births is no longer increasing. If extreme poverty keeps falling, and sex education and contraception keep spreading, then the world population will keep growing fast, but only until the inevitable fill-up is completed.

Wait, "They" Still Have Many Children

Even after I show these charts onstage, people come up to me after the presentation and tell me that the charts can't be correct because, you know, *"People in Africa and Latin America still have many children. And religious people refuse contraceptives and still have huge families."*

Skilled journalists pick and choose dramatic exceptional people in their reports. In the mass media we sometimes see examples of very religious people, whether living in traditional ways or leading seemingly modern lives, who proudly show us their very large families as evidence of faith. Such documentary films, TV shows, and media reports give the impression that religion leads to much larger families. But whatever their religion—whether they are Catholics, Jews, or Muslims—these families share one quality. They are the exceptions!

In reality, the connection between religion and babies per woman is not so impressive. Throughout this book I discuss how the media chooses its exceptional stories, and in chapter 7 I will debunk the myth of religion and large families. For now, let's look at the single factor that does have a strong connection with large families: extreme poverty.

Why More Survivors Lead to Fewer People

When combining all the parents living on Levels 2, 3, and 4, from every region of the world, and of every religion or no religion, together they have on average two children. No kidding! This includes the populations of Iran, Mexico, India, Tunisia, Bangladesh, Brazil, Turkey, Indonesia, and Sri Lanka, just to name a few examples.

The poorest 10 percent combined still have five children on average. And on average, every second family living in extreme poverty loses one of their children before he or she reaches the age of five. That is shamefully high, but still far better than the ghastly levels that kept population growth down in the bad old times.

AVERAGE FAMILY SIZE BY INCOME, 2017

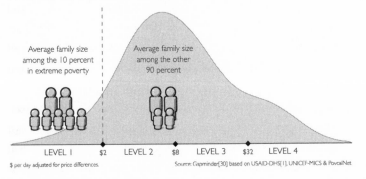

Average family size among the 10 percent in extreme poverty

Average family size among the other 90 percent

LEVEL 1 $2   LEVEL 2 $8   LEVEL 3 $32   LEVEL 4

$ per day adjusted for price differences.   Source: Gapminder[30] based on USAID-DHS[1], UNICEF-MICS & PovcalNet

When people hear that the population is growing, they intuitively think it will continue to grow unless something is done. They intuitively visualize the trend continuing into the future. But remember, for my grandchild Mino to stop growing taller, nothing drastic needs to be done.

Melinda Gates runs a philanthropic foundation together with her husband, Bill. They have spent billions of dollars to save the lives of millions of children in extreme poverty by investing in primary health care and education. Yet intelligent and well-meaning people keep contacting their foundation saying that they should stop. The argument goes like this: *"If you keep saving poor children, you'll kill the planet by causing overpopulation."*

I have also heard this argument after some of my presentations, from people who may have the best intentions and want to save the planet for future generations. It sounds intuitively correct. If more children survive, the population *just* increases. Right? No! Completely wrong.

Parents in extreme poverty need many children for the reasons I set out earlier: for child labor but also to have extra children in case some children die. It is the countries with the highest child mortality rates, like Somalia, Chad, Mali, and Niger, where women have the most babies: between five and eight. Once parents see children survive, once the children are no longer needed for child labor, and once the women are educated and have information about and access to contraceptives, across cultures and religions both the men and the women instead start dreaming of having fewer, well-educated children.

"Saving poor children *just* increases the population" sounds correct, but the opposite is true. Delaying the escape from extreme poverty *just* increases the population. Every generation kept in extreme poverty will produce an even larger next generation. The only proven method for curbing population growth is to eradicate extreme poverty and give people better lives, including education and contraceptives. Across the world, parents then have chosen for themselves to have fewer children. This transformation has happened across the world but it has never happened without lowering child mortality.

This discussion so far has left out the most important point, which is the moral imperative to help people escape from the misery and indignity of extreme poverty. The argument that we must save the planet for future people, not yet born, is difficult for me to hear when people are suffering today. But when it comes to child mortality, we don't have to choose between the present and the future, or between our hearts and our heads: they all point in the same direction. We should do everything we can to reduce child mortality, not only as an act of humanity for living suffering children but to benefit the whole world now and in the future.

***

Two Public Health Miracles

In the first full year of Bangladesh's independence, 1972, Bangladeshi women had on average seven children and life expectancy was 52. Today, Bangladeshi women have two children and a newborn can expect to live for 73 years. In four decades, Bangladesh has gone from miserable to decent. From Level 1 to Level 2. It is a miracle, delivered through remarkable progress in basic health and child survival. The child survival rate is now 97 percent—up from less than 80 percent at independence. Now that parents have reason to expect that all their children will survive, a major reason for having big families is gone.

In Egypt in 1960, 30 percent of all children died before their fifth birthday. The Nile delta was a misery for children, with all sorts of dangerous diseases and malnutrition. Then a miracle happened. The Egyptians built the Aswan Dam, they wired electricity into people's homes, improved education, built up primary health care, eradicated malaria, and made drinking water safer. Today, Egypt's child mortality rate, at 2.3 percent, is lower than it was in France or the United Kingdom in 1960.

***

How to Control the Straight Line Instinct, or Not All Lines Are Straight

The best way of controlling the instinct to always see straight lines—whether in relation to population growth or in other situations—is simply to remember that curves naturally come in lots of different shapes. Many aspects of the world are best represented by curves shaped like an S, or a slide, or a hump, and not by a straight line. Here are some examples, each showing how a particular aspect of life changes as we move across the four income levels.

Straight Lines

Straight lines are much less common than we tend to think, but some lines are straight. Below is a simplified version of the wealth and health chart you have seen before. Instead of all the bubbles, we can draw a line where most of the bubbles are. Some bubbles are above the line and others are below but you can see that in general they cluster around a straight line.

A STRAIGHT LINE
Longer lives and higher income go hand in hand.

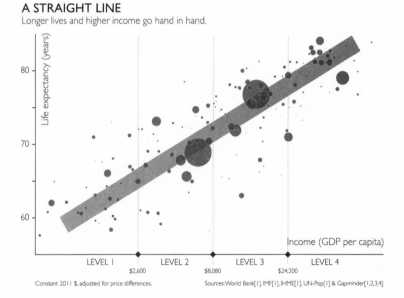

Constant 2011 $, adjusted for price differences.

Sources: World Bank[1], IMF[1], IHME[1], UN-Pop[1] & Gapminder[1,2,3,4]

This chart shows that money and health go hand in hand. We don't know from just looking at the line which comes first or what the relationship is between the two. It might be that a healthy population produces more income. It might be that a rich population can afford better health. I think both are true. What we do know from such a line is that in general where income is higher, health is better.

We can also find straight lines when we compare income levels with education, marriage age, and spending on recreation. More income goes hand in hand with longer average schooling, with women marrying later, and with a greater share of income going toward recreation.

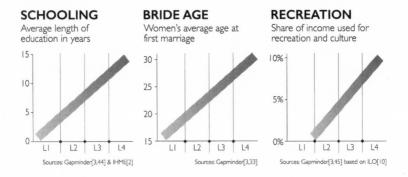

SCHOOLING
Average length of education in years

BRIDE AGE
Women's average age at first marriage

RECREATION
Share of income used for recreation and culture

Sources: Gapminder[3,44] & IHME[2]

Sources: Gapminder[3,33]

Sources: Gapminder[3,45] based on ILO[10]

S-Bends

When we compare income with basic necessities like primary-level education or vaccination, we see S-shaped curves. They are low and flat at Level 1, then they rise quickly through Level 2, because above Level 1, countries can afford primary education and vaccination (the most cost-effective health intervention there is) for just about the entire population. Just as we will buy ourselves a fridge and a cell phone as soon as we can afford them, countries will invest in primary education and vaccination as soon as they can afford them. Then the curves flatten off at Levels 3 and 4. Everyone already has these things. The curves reach their maximum and stay there.

Remembering about this kind of curve will help you to improve your guessing about the world: on Level 2, almost everyone can already afford to have their basic physical needs met.

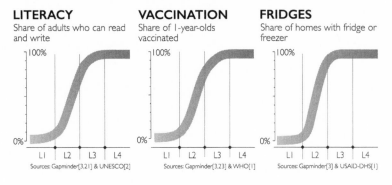

LITERACY
Share of adults who can read and write

VACCINATION
Share of 1-year-olds vaccinated

FRIDGES
Share of homes with fridge or freezer

Sources: Gapminder[3,21] & UNESCO[2]　　Sources: Gapminder[3,23] & WHO[1]　　Sources: Gapminder[3] & USAID-DHS[1]

Slides

The babies-per-woman curve looks like a slide in a playground. It starts flat, then, after a certain level of income, it slopes downward, and then it flattens out and stays quite low, just below two babies per woman.

A SLIDE
In this graph dots may represent countries, or, wherever we had data, we split a country into five income groups, each representing 20 percent of the population. This shows 2017.

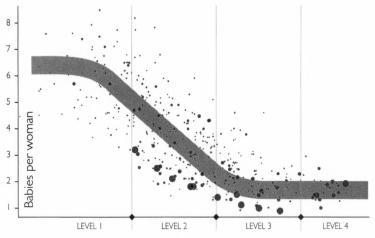

Source: Gapminder[3,47] based on GDL[1], USAID-DHS[1], UNICEF-MICS & OurWorldInData[10]

Shifting away from income graphs for a moment, we see a similar shape for the cost of vaccinations. In basic math classes, we teach children to multiply. If an injection costs $10, what's the price of a million injections? UNICEF knows how to count but it has also saved millions of children's lives by not accepting a straight line. It has negotiated huge contracts with pharmaceutical companies, in which the price is cut to the bare minimum in return for guaranteed long contracts. But when you have negotiated to the bottom price, you can't get lower. That's another slide-shaped curve.

Humps

Your tomato plant will grow as long as it gets water. So, if more water is what it needs, why don't you turn the hose on it, so you can grow an enormous prize-winning tomato? Of course you know that doesn't work. It's a question of dosage. Too little and it dies. Too much and it dies too. Tomato survival is low in very dry and very wet environments, but high in environments that are in the middle.

Similarly, there are some phenomena that are lower in countries on Level 1 and countries on Level 4, but higher in middle-income countries—which means the majority of countries.

Dental health, for example, gets worse as people move from Level 1 to Level 2, then improves again on Level 4. This is because people start to eat sweets as soon as they can afford them, but their governments cannot afford to prioritize preventive public education about tooth decay until Level 3. So poor teeth are an indicator of relative poverty on Level 4, but on Level 1 they may indicate the opposite.

Motor vehicle accidents show a similar hump-shaped pattern. Countries on Level 1 have fewer motor vehicles per person, so they do not have many motor vehicle accidents. In countries on Levels 2 and

3, the poorest people keep walking the roads while others start to travel by motor vehicles—minibuses and motorcycles—but roads, traffic regulations, and traffic education are still poor, so accidents reach a peak, before they decline again in countries on Level 4. The same goes for child drownings as a percentage of all child deaths.

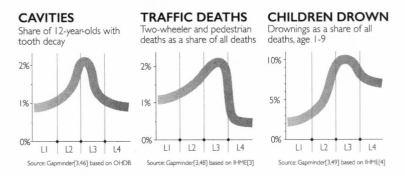

CAVITIES
Share of 12-year-olds with tooth decay

TRAFFIC DEATHS
Two-wheeler and pedestrian deaths as a share of all deaths

CHILDREN DROWN
Drownings as a share of all deaths, age 1-9

Source: Gapminder[3,46] based on OHDB   Source: Gapminder[3,48] based on IHME[3]   Source: Gapminder[3,49] based on IHME[4]

Like tomatoes, human beings need water to survive. But if you drink six liters at once, you will die. The same goes for sugar, fat, and medicines. Actually, everything you need to survive is lethal in high dosage. Too much stress is bad, but the right amount improves performance. Self-confidence has its optimal dosage. The intake of dramatic news from the rest of the world probably has its optimal dosage too.

Doubling Lines

Finally, doubling. The doubling pattern of the Ebola virus is actually a very common type of pattern in nature. For example, the number of *E. coli* bacteria in a body can explode in just a few days because it can double every 12 hours: 1, 2, 4, 8, 16, 32 . . . The world of transport also contains many doubling patterns. As people's incomes increase, the distance they travel each year keeps doubling. So does the share of their income that they spend on transport. On Level 4, transport is behind one-third of all CO_2 emissions—which also double with income.

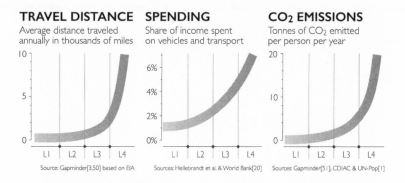

TRAVEL DISTANCE
Average distance traveled annually in thousands of miles

Source: Gapminder[3,50] based on EIA

SPENDING
Share of income spent on vehicles and transport

Sources: Hellebrandt et al. & World Bank[20]

CO₂ EMISSIONS
Tonnes of CO_2 emitted per person per year

Sources: Gapminder[51], CDIAC & UN-Pop[1]

Most people's incomes grow much slower than bacteria, unfortunately. Still, even if your income increases by only 2 percent a year, after 35 years it will have doubled. And then, if you maintain 2 percent growth, in another 35 years it will have doubled again. Over 200 years—if you lived that long—it would double six times, which is exactly what we saw in Sweden's bubble trail in the last chapter, and which is typically the slow and steady way countries have moved from Level 1 to Level 4. The graph on the next page shows how six doublings move you across all four income levels.

I have divided the levels in this way because that's how money works. The impact of an additional dollar is not the same on different levels. On Level 1, with $1 a day, another dollar buys you that extra bucket. That is life-changing. On Level 4, with $64 a day, another dollar has almost no impact. But with another $64 a day, you could build a pool or buy a summer house. That's life-changing for you. The world is extremely unfair, but doubling one's income, from any starting point, is always life-changing. I use this doubling scale whenever I compare income because that's how money works.

By the way, the scales for measuring earthquakes, sound levels, and pH works in the same way.

DOUBLING INCOME

Daily income doubles twice from one level to the next

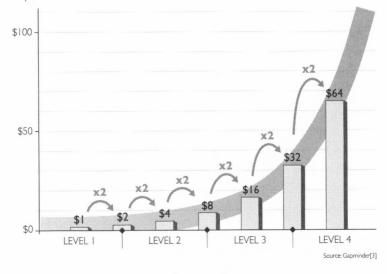

Source: Gapminder[3]

How Much of the Curve Do You See?

Curves come in many different shapes. The part of the curve with which we are familiar, living on Level 4, may not apply at all on Levels 1, 2, or 3. An apparently straight upward trend could be part of a straight line, an S-bend, a hump, or a doubling line. An apparently straight downward trend could be part of a straight line, a slide, or a hump. Any two connected points look like a straight line but when we have three points we can distinguish between a straight line (1, 2, 3) and the start of what may be a doubling line (1, 2, 4).

To understand a phenomenon, we need to make sure we understand the shape of its curve. By assuming we know how a curve continues beyond what we see, we will draw the wrong conclusions and come up with the wrong solutions. That is what I did before I realized that the Ebola epidemic was doubling. And that is what everyone is doing who thinks that the world population is _just_ increasing.

Factfulness

Factfulness is . . . recognizing the assumption that a line will just continue straight, and remembering that such lines are rare in reality.

To control the straight line instinct, **remember that curves come in different shapes.**

- **Don't assume straight lines.** Many trends do not follow straight lines but are S-bends, slides, humps, or doubling lines. No child ever kept up the rate of growth it achieved in its first six months, and no parents would expect it to.

CHAPTER FOUR

THE FEAR INSTINCT

How to hide 40 million airplanes, and how
I kind of won the Nobel Peace Prize

Blood All Over the Floor

On October 7, 1975, I was plastering a patient's arm when an assistant nurse burst through the door and announced that a plane had crashed and the wounded were coming in by helicopter. It was my fifth day as a junior doctor on the emergency ward in the small coastal town of Hudiksvall in Sweden. All the senior staff were down in the dining hall and as the assistant nurse and I searched frantically for the folder of disaster instructions, I could already hear the helicopter landing. The two of us were going to have to handle this on our own.

Seconds later a stretcher was rolled in, bearing a man in dark green overalls and a camouflage life jacket. His arms and legs were twitching. An epileptic seizure, I thought; off with his clothes. I removed his life jacket easily but his overalls were more problematic. They looked like a spacesuit, with huge sturdy zippers all over, and no matter how hard I tried I couldn't find the zipper that undid them. I had just registered that the uniform meant this was a military pilot when I noticed the blood all over the floor. "He's bleeding," I shouted. With this much blood, I knew he could be dead in a matter of seconds, but with the overalls still on, I couldn't see where it was coming from. I grabbed a big pair of plaster pliers to cut through the fabric and howled to the assistant nurse, "Four bags of blood, O-negative. Now!"

To the patient, I shouted, "Where does it hurt?" "Yazhe shisha . . . na adjezhizha zha . . ." he replied. I couldn't understand a word, but it sounded like Russian. I looked the man in his eyes and said with a clear voice, "все тихо товарищ, шведскауа больница," which means "All is calm, comrade, Swedish hospital."

I will never forget the look of panic I triggered with those words. Frightened out of his mind, he stared back at me and tried to tell me something: "Vavdvfor papratarjenji rysskamememje ej . . ." I looked into his eyes full of fear, and then I realized: this must be a Russian fighter pilot who has been shot down over Swedish territory. Which means that the Soviet Union is attacking us. World War III has started! I was paralyzed by fear.

Fortunately, at that moment the head nurse, Birgitta, came back from lunch. She snatched the plaster pliers from my hand and hissed, "Don't shred it. That's an air force 'G suit' and it costs more than 10,000 Swedish kronor." After a beat she added, "And can you please step off the life jacket. You're standing on the color cartridge and it is making the whole floor red."

Birgitta turned to the patient, calmly freed him from his G suit, and wrapped him in a couple of blankets. In the meantime she told him in Swedish, "You were in the icy water for 23 minutes, which is why you are jerking and shivering, and why we can't understand what you're saying." The Swedish air force pilot, who had evidently crashed during a routine flight, gave me a comforting little smile.

A few years ago I contacted the pilot, and was relieved to hear that he doesn't remember a thing from those first minutes in the emergency room in 1975. But for me the experience is hard to forget. I will forever remember my complete misjudgment. Everything was the other way around: the Russian was Swedish, the war was peace, the epileptic seizure was cooling, and the blood was a color ampule from inside the life jacket. Yet it had all seemed so convincing to me.

When we are afraid, we do not see clearly. I was a young doctor facing my first emergency, and I had always been terrified by the prospect of a third world war. As a child, I often had nightmares about it. I would wake up and run to my parents' bed. I could be calmed only by my father going over the details of our plan one more time: we would take our tent in the bike trailer and go live in the woods where there were plenty of blueberries. Inexperienced, and in an emergency situation for the first time, my head quickly generated a worst-case scenario. I didn't see what I wanted to see. I saw what I was afraid of seeing. Critical thinking is always difficult, but it's almost impossible when we are scared. There's no room for facts when our minds are occupied by fear.

The Attention Filter

None of us has enough mental capacity to consume all the information out there. The question is, what part are we processing and how did it get selected? And what part are we ignoring? The kind of information we seem most likely to process is stories: information that sounds dramatic.

Imagine that we have a shield, or attention filter, between the world and our brain. This attention filter protects us against the noise of the world: without it, we would constantly be bombarded with so much information we would be overloaded and paralyzed. Then imagine that the attention filter has ten instinct-shaped holes in it—gap, negativity, straight line, and so on. Most information doesn't get through, but the holes do allow through information that appeals to our dramatic instincts. So we end up paying attention to information that fits our dramatic instincts, and ignoring information that does not.

The media can't waste time on stories that won't pass our attention filters.

Here are a couple of headlines that won't get past a newspaper editor, because they are unlikely to get past our own filters: "MALARIA CONTINUES TO GRADUALLY DECLINE." "METEOROLOGISTS CORRECTLY PREDICTED YESTERDAY THAT THERE WOULD BE MILD WEATHER IN LONDON TODAY." Here are some topics that easily get through our filters: earthquakes, war, refugees, disease, fire, floods, shark attacks, terror attacks. These unusual events are more newsworthy than everyday ones. And the unusual stories we are constantly shown by the media paint pictures in our heads. If we are not extremely careful, we come to believe that the unusual is usual: that this is what the world looks like.

For the first time in world history, data exists for almost every aspect of global development. And yet, because of our dramatic instincts and the way the media must tap into them to grab our attention, we continue to have an overdramatic worldview. Of all our dramatic instincts, it seems to be the fear instinct that most strongly influences what information gets selected by news producers and presented to us consumers.

The Fear Instinct

When people are asked in polls what they are most afraid of, four answers always tend to turn up near the top: snakes, spiders, heights, and being trapped in small spaces. Then comes a long list with no surprises: public speaking, needles, airplanes, mice, strangers, dogs, crowds, blood, darkness, fire, drowning, and so on.

These fears are hardwired deep in our brains for obvious evolutionary reasons. Fears of physical harm, captivity, and poison once helped our ancestors survive. In modern times, perceptions of these dangers still trigger our fear instinct. You can spot stories about them in the news every day:

- physical harm: violence caused by people, animals, sharp objects, or forces of nature
- captivity: entrapment, loss of control, or loss of freedom
- contamination: by invisible substances that can infect or poison us

These fears are still constructive for people on Levels 1 and 2. For example, it is practical, on Levels 1 and 2, to be afraid of snakes. Sixty thousand people are killed by snakes every year. Better to jump one too many times when you see a stick. Whatever you do, don't get bitten. There's no hospital nearby and if there is you can't afford it.

A Midwife's Wish

In 1999, I traveled with a couple of Swedish students to visit a traditional midwife in a remote village in Tanzania. I wanted my medical students from Level 4 to meet a real health worker on Level 1 instead of just reading about them in books. The midwife had no formal education, and the students' jaws dropped when she described her struggles, walking between villages to help poor women deliver babies on mud floors in complete darkness with no medical equipment and no clean water.

One of the students asked, "Do you have children of your own?" "Yes," she said proudly, "two boys and two daughters." "Will your daughters become midwives like you?" The old woman threw her body forward and laughed out loud. "My daughters! Working like me?! Oh no! Never! Ever! They have nice jobs. They work in front of computers in Dar es Salaam, just like they wanted to." The midwife's daughters had escaped Level 1.

Another student asked, "If you could choose one piece of equipment that could make your work easier, what would that be?" "I really want a flashlight," she answered. "When I walk to a village in the dark, even when the moon is shining, it is so difficult to see the snakes."

On Levels 3 and 4, where life is less physically demanding and people can afford to protect themselves against nature, these biological memories probably cause more harm than good. On Level 4, for sure the fears that evolved to protect us are now doing us harm. A small minority—3 percent—of the population on Level 4 suffers from a phobia so strong it hinders their daily life. For the vast majority of us not blocked by phobias, the fear instinct harms us by distorting our worldview.

The media cannot resist tapping into our fear instinct. It is such an easy way to grab our attention. In fact the biggest stories are often those that trigger more than one type of fear. Kidnappings and plane crashes, for example, each combine the fear of harm and the fear of captivity. Earthquake victims trapped under collapsed buildings are both hurt and trapped, and get more attention than regular earthquake victims. The drama is so much stronger when multiple fears are triggered.

Yet here's the paradox: the image of a dangerous world has never been broadcast more effectively than it is now, while the world has never been less violent and more safe.

Fears that once helped keep our ancestors alive, today help keep journalists employed. It isn't the journalists' fault and we shouldn't expect them to change. It isn't driven by "media logic" among the producers so much as by "attention logic" in the heads of the consumers.

If we look at the facts behind the headlines, we can see how the fear instinct systematically distorts what we see of the world.

Natural Disasters: In Times Like These

Nepal is one of the last Asian countries left on Level 1, and in 2015 it was hit by an earthquake. The death rate is always higher when a disaster hits a country on Level 1, because of poorly constructed buildings, poor infrastructure, and poor medical facilities. Nine thousand people died.

FACT QUESTION 7

How did the number of deaths per year from natural disasters change over the last hundred years?

☐ A: More than doubled
☐ B: Remained about the same
☐ C: Decreased to less than half

This number includes all fatalities from floods, earthquakes, storms, droughts, wildfires, and extreme temperatures, and also deaths during the mass displacement of people and pandemics after such events. Just 10 percent of people picked the right answer, and even in the countries that did best on this question—Finland and Norway— it was only 16 percent. (As always, the full country breakdown is in the appendix.) The chimpanzees, who don't watch the news, got 33 percent as always! In fact, the number of deaths from acts of nature has dropped far below half. It is now just 25 percent of what it was 100 years ago. The human population increased by 5 billion people over the same period, so the drop in deaths per capita is even more amazing. It has fallen to just 6 percent of what it was 100 years ago.

The reason natural disasters kill so many fewer people today is not that nature has changed. It is that the majority of people no longer live on Level 1. Disasters hit countries on all income levels, but the harm done is very different. With more money comes better preparedness. The graph below shows the average number of deaths from natural disasters per million people over the last 25 years, on each income level.

DISASTER PROTECTION COSTS MONEY

Annual deaths from natural disasters, per million people.
Average of the 25-year period 1991-2016.

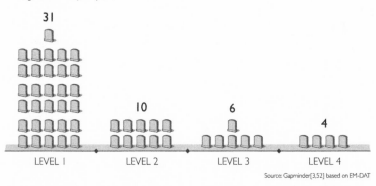

Source: Gapminder[3,52] based on EM-DAT

Thanks to better education, new affordable solutions, and global collaborations, the decrease in death rates is impressive even among those who are stuck on Level 1—as shown in the image on the next page.

(We look at averages of 25-year periods because natural disasters don't occur at an even rate each year. Even so, just one event, the heat wave in Europe in 2003, was largely responsible for the fourfold increase in the death rate on Level 4.)

DISASTERS KILL FEWER ON LEVEL I
Average annual deaths from natural disasters, per million people.

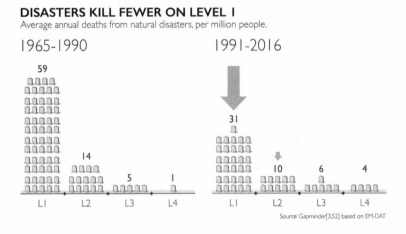

1965-1990 — 1991-2016

Source: Gapminder[3,52] based on EM-DAT

Back in 1942, Bangladesh was on Level 1 and almost all its citizens were illiterate farmers. Over a two-year period it suffered terrible floods, droughts, and cyclones. No international organization came to the rescue and 2 million people died. Today, Bangladesh is on Level 2. Today, almost all Bangladeshi children finish school, where they learn that three red-and-black flags means everyone must run to the evacuation centers. Today, the government has installed across the country's huge river delta a digital surveillance system connected to a freely available flood-monitoring website. Just 15 years ago, no country in the world had such an advanced system. When another cyclone hit in 2015, the plan worked and the World Food Programme flew in 113 tons of high-energy biscuits to the 30,000 evacuated families.

In the same year, vivid images spread awareness across the world of the horrific earthquake in Nepal, and rescue teams and helicopters were quickly deployed. Tragically, thousands were already dead, but the

humanitarian resources that rushed to this inaccessible country on
Level 1 did manage to prevent the death toll from rising even further.

The UN's ReliefWeb has become a global coordinator for disaster
help—something earlier generations of disaster victims could only
dream of. And it is paid for by taxpayers on Level 4. We should be
very proud of it. We humans have finally figured out how to protect
ourselves against nature. The huge reduction in deaths from natural
disasters is yet another trend to add to the pile of mankind's ignored,
unknown success stories.

Unfortunately, the people on Level 4 paying for ReliefWeb are the
same people we asked about the trend in natural disasters. Ninety-
one percent of them are unaware of the success they are paying for
because their journalists continue to report every disaster as if it were
the worst. The long, elegantly dropping trend line, a bit of fact-based
hope, they think is not newsworthy.

DEATHS FROM DISASTER
Annual deaths per million people, 10-year averages.

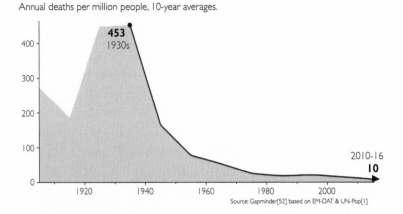

Source: Gapminder[52] based on EM-DAT & UN-Pop[1]

Next time the news shows you horrific images of victims trapped
under collapsed buildings, will you be able to remember the positive
long-term trend? When the journalist turns to the camera and says,
"The world just became a bit more dangerous," will you be able to dis-

agree? To look at the local rescue crew in their colorful helmets and think, "Most of their parents couldn't read. But these guys are following internationally used first-aid guidelines. The world is getting better."

When the journalist says with a sad face, "in times like these," will you smile and think that she is referring to the first time in history when disaster victims get immediate global attention and foreigners send their best helicopters? Will you feel fact-based hope that humanity will be able to prevent even more horrific deaths in the future?

I don't think so. Not if you function like me. Because when that camera pans to bodies of dead children being pulled out of the debris, my intellectual capacity is blocked by fear and sorrow. At that moment, no line chart in the world can influence my feelings, no facts can comfort me. Claiming in that moment that things are getting better would be to trivialize the immense suffering of those victims and their families. It would be absolutely unethical. In these situations we must forget the big picture and do everything we can to help.

The big facts and the big picture must wait until the danger is over. But then we must dare to establish a fact-based worldview again. We must cool our brains and compare the numbers to make sure our resources are used effectively to stop future suffering. We can't let fear guide these priorities. Because the risks we fear the most are now often—thanks to our successful international collaboration—the risks that actually cause us the least harm.

For ten days or so in 2015 the world was watching the images from Nepal, where 9,000 people had died. During the same ten days, diarrhea from contaminated drinking water also killed 9,000 children across the world. There were no camera teams around as these children fainted in the arms of their crying parents. No cool helicopters swooped in. Helicopters, anyway, don't work against this child killer (one of the world's worst). All that's needed to stop a child from accidentally drinking her neighbor's still-lukewarm poo is a few plastic pipes, a water pump, some soap, and a basic sewage system. Much cheaper than a helicopter.

40 Million Invisible Planes

In 2016 a total of 40 million commercial passenger flights landed safely at their destinations. Only ten ended in fatal accidents. Of course, those were the ones the journalists wrote about: 0.000025 percent of the total. Safe flights are not newsworthy. Imagine:

"Flight BA0016 from Sydney arrived in Singapore Changi airport without any problems. And that was today's news."

2016 was the second safest year in aviation history. That is not newsworthy either.

This graph shows plane crash deaths per 10 billion commercial passenger miles over the last 70 years. Flying has gotten 2,100 times safer.

PLANE CRASH DEATHS
Annual deaths per 10 billion passenger miles, by commercial airlines. Five-year averages.

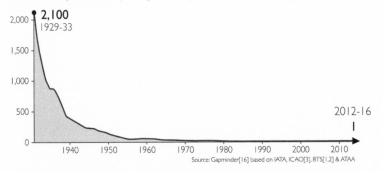

Source: Gapminder[16] based on IATA, ICAO[3], BTS[1,2] & ATAA

Back in the 1930s, flying was really dangerous and passengers were scared away by the many accidents. Flight authorities across the world had understood the potential of commercial passenger air traffic, but they also realized flying had to become safer before most people would dare to try it. In 1944 they all met in Chicago to agree on common rules and signed a contract with a very important Annex 13: a common form for incident reports, which they agreed to share, so they could all learn from each other's mistakes.

Since then, every crash or incident involving a commercial passenger airplane has been investigated and reported; risk factors have been systematically identified; and improved safety procedures have been adopted, worldwide. Wow! I'd say the Chicago Convention is one of humanity's most impressive collaborations ever. It's amazing how well people can work together when they share the same fears.

The fear instinct is so strong that it can make people collaborate across the world, to make the greatest progress. It's so strong it can also remove 40 million noncrashing aircraft from our field of sight each year. Just like it can erase 330,000 child deaths from diarrhea from our TV screens. Just like that.

War and Conflict

I was born in 1948, three years after the end of the Second World War, in which 65 million people died. No one pretended that another world war could not come. And yet, it did not come. Instead came peace: the longest peace between superpowers in human history.

Today, conflicts and fatalities from conflicts are at a record low. I have lived through the most peaceful decades in human history. Watching the news, with its never-ending flow of horrifying images, it is almost impossible to believe that.

I do not seek to trivialize the horror that undoubtedly remains. I do not try to understate the importance of ending current conflicts. Remember: things can be bad, *and* getting better. Getting better, but still bad. The world was once mostly barbaric and it is now mostly not. But for the people of Syria, these trends are of course not comforting. There it is barbaric right now.

The Syrian conflict will most likely prove to be the deadliest in the world since the Ethiopian-Eritrean war of 1998 to 2000. We don't know the total fatalities yet and we don't know if the conflict will spread. If fatalities end up being in the tens of thousands, the conflict will have been less bloody than the worst wars of the 1990s. If the death toll reaches 200,000, this will still fall short of the wars of the 1980s. This is no comfort whatsoever to those living through this horror, but the fact that battle deaths are falling decade by decade should be some comfort to the rest of us.

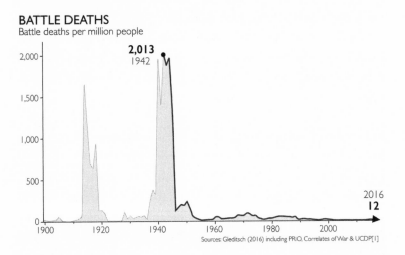

BATTLE DEATHS
Battle deaths per million people

Sources: Gleditsch (2016) including PRIO, Correlates of War & UCDP[1]

The general trend toward less violence is not just one more improvement. It is the most beautiful trend there is. The spread of peace over the last decades has enabled all the other improvements we have seen. We must take care of this fragile gift if we hope to achieve our other noble goals, such as collaboration toward a sustainable future. Without world peace, you can forget about all other global progress.

Contamination

The threat of a third, nuclear, world war was very real to me during my childhood in the 1950s and throughout the next three decades. It was real to most people. We all had images in our heads of the victims of Hiroshima, and the news showed superpowers flexing their nuclear muscles like bodybuilders on steroids, one test bombing after another. In 1985, the Nobel Peace Prize committee decided that nuclear disarmament was the most important peace cause in the world. They awarded the prize to me. Well, not to me directly, but to IPPNW, the International Physicians for the Prevention of Nuclear War, and I was a proud member of that organization.

In 1986 there were 64,000 nuclear warheads in the world; today there are 15,000. So the fear instinct can sure help to remove terrible things from the world. On other occasions, it runs out of control, distorts our risk assessment, and causes terrible harm.

Eighteen miles underwater, on the seafloor of the Pacific Ocean just off the coast of Japan, a "seismic slip-rupture event" took place on March 11, 2011. It moved the Japanese main island eight feet eastward and generated a tsunami that reached the coast one hour later, killing roughly 18,000 people. The tsunami also was higher than the wall that was built to protect the nuclear power plant in Fukushima. The province was flooded with water and the world's news was flooded with fear of physical harm and radioactive contamination.

People escaped the province as fast as they could, but 1,600 more people died. It was not the leaking radioactivity that killed them. Not one person has yet been reported as having died from the very thing that people were fleeing from. These 1,600 people died because they escaped. They were mainly old people who died because of the mental and physical stresses of the evacuation itself or of life in evacuation shelters. It wasn't radioactivity, but the fear of radioactivity, that killed them. (Even after the worst-ever nuclear accident, Chernobyl in 1986, when most people expected a huge increase in the death rate, the WHO investigators could not confirm this, even among those living in the area.)

In the 1940s, a new wonder chemical was discovered that killed many annoying insects. Farmers were so happy. People fighting malaria were so happy. DDT was sprayed over crops, across swamps, and in homes with little investigation of its side effects. DDT's creator won a Nobel Prize.

During the 1950s the early environmental movement in the United States started to raise concerns about levels of DDT accumulating up the food chain into fish and even birds. The great popular science writer Rachel Carson reported her finding that the shells of bird eggs in her area were becoming thinner in *Silent Spring*, a book that became a global bestseller. The idea that humans were allowed to spread invisible substances to kill bugs, and authorities were looking away from any signs of the wider impact on other animals or on humans, was of course frightening.

A fear of insufficient regulation and of irresponsible companies was ignited and the global environmental movement was born. Thanks to this movement—and to further contamination scandals involving oil spills, plantation workers disabled by pesticides, nuclear reactor failures—the world today has decent chemical and safety regulations covering many countries (though still not close to the impressive coverage of the aviation industry). DDT was banned in several countries and aid agencies had to stop using it.

But. *But.* As a side effect, we have been left with a level of public fear of chemical contamination that almost resembles paranoia. It is called chemophobia.

This means that a fact-based understanding of topics like childhood vaccinations, nuclear power, and DDT is still extremely difficult today. The memory of insufficient regulation has created automatic mistrust and fear, which blocks the ability to hear data-driven arguments. I will try anyway.

In a devastating example of critical thinking gone bad, highly educated, deeply caring parents avoid the vaccinations that would protect their children from killer diseases. I love critical thinking and I admire skepticism, but only within a framework that respects the evidence. So if you are skeptical about the measles vaccination, I ask you to do two things. First, make sure you know what it looks like when a child dies from measles. Most children who catch measles recover, but there is still no cure and even with the best modern medicine, one or two in every thousand will die of it. Second, ask yourself, "What kind of evidence would convince me to change my mind?" If the answer is "no evidence could ever change my mind about vaccination," then you are putting yourself outside evidence-based rationality, outside the very critical thinking that first brought you to this point. In that case, to be consistent in your skepticism about science, next time you have an operation please ask your surgeon not to bother washing her hands.

More than one thousand old people died escaping from a nuclear leak that killed no one. DDT is harmful but I have been unable to find numbers showing that it has directly killed anyone either. The harm investigations that were not done in the 1940s have been done now. In 2002 the Centers for Disease Control and Prevention produced a 497-page document named *Toxicological Profile for DDT, DDE and DDD*. In 2006 the World Health Organization finally finished reviewing all the scientific investigations and, just like the CDC, classified DDT as "mildly harmful" to humans, stating that it had more health benefits than drawbacks in many situations.

DDT should be used with great caution, but there are pros and cons. In refugee camps teeming with mosquitoes, for example, DDT is often one of the quickest and cheapest ways to save lives. Americans, Europeans, and fear-driven lobbyists, though, refuse to read the CDC's and WHO's lengthy investigations and short recommendations and are not ready to discuss the use of DDT. Which means some aid organizations that depend on popular support avoid evidence-based solutions that actually would save lives.

Improvements in regulations have been driven not by death rates but by fear, and in some cases—Fukushima, DDT—fear of an invisible substance has run amok and is doing more harm than the substance is itself.

The environment is deteriorating in many parts of the world. But just as dramatic earthquakes receive more news coverage than diarrhea, small but scary chemical contaminations receive more news coverage than more harmful but less dramatic environmental deteriorations, such as the dying seabed and the urgent matter of overfishing.

Chemophobia also means that every six months there is a "new scientific finding" about a synthetic chemical found in regular food in very low quantities that, if you ate a cargo ship or two of it every day for three years, could kill you. At this point, highly educated people put on their worried faces and discuss it over a glass of red wine. The zero-death toll seems to be of no interest in these discussions. The level of fear seems entirely driven by the "chemical" nature of the invisible substance.

Now let's move to the latest number one fear in the West.

Terrorism

If there's one group of people who have fully understood the power of the fear instinct, it's not journalists. It's terrorists. The clue is in their name. Fear is what they aim for. And they succeed by tapping into all our primitive fears—of physical harm, of being trapped, of being poisoned or contaminated.

Terrorism is one of the exceptions to the global trends discussed in chapter 2 on negativity. It is getting worse. So are you right to be very scared of it? Well, first of all it accounted for 0.05 percent of all deaths in the world in 2016, so probably not. Second, it depends where you live.

At the University of Maryland in the United States, a group of researchers has collected data about all terror events recorded in reliable media since 1970. The result is the freely available Global Terrorism Database, containing details of 170,000 terror events. This database shows that in the ten-year period from 2007 to 2016, terrorists killed 159,000 people worldwide: three times more than the number killed in the previous ten-year period. Just like with Ebola, when a number is doubling or tripling, of course we should be worried and look closer to see what it means.

Hunting Terrorism Data

In this part of the book, all the trends finish in 2016 because 2016 is the last year of data in the Global Terrorism Database. The researchers carefully study multiple sources to eliminate rumors and false information for each record they enter, which creates a time delay. That is good scientific practice, but I find it strange. Just like with Ebola, and as with the CO_2 emissions I will discuss later, when something seems important and concerning, don't we need up-to-date data as quickly as possible rather than perfect data? Otherwise how can we know whether terrorism is increasing or not?

Wikipedia contains articles with long lists of recent terror attacks from all over the world. Volunteers update them amazingly quickly, just minutes after the first news is out. I love Wikipedia and if we could trust these lists, we wouldn't have to wait so long to see the trend. To check their reliability we decided to compare (English) Wikipedia with the Global Terrorism Database for 2015. If the overlap was close to 100 percent, we could probably trust Wikipedia to be quite complete for 2016 and 2017 as well, and use it as a good-enough source for tracking more up-to-date terrorism trends.

It turned out Wikipedia unintentionally presented a very distorted worldview. It was distorted in a systematic way according to a Western mind-set. Our disappointment was huge. More precisely, it was 78 percent. That's how many of the 2015 terrorism deaths were missing from Wikipedia. While almost all the deaths in the West were recorded, only 25 percent of those in "the rest" were there.

No matter how much I love Wikipedia, we still need serious researchers to maintain reliable data sets. But they need more resources so they can update them quicker.

FEWER TERRORISM DEATHS ON LEVEL 4

Globally, terrorists killed three times more people than in the previous 10-year period.
Each grave represents 1,000 deaths.

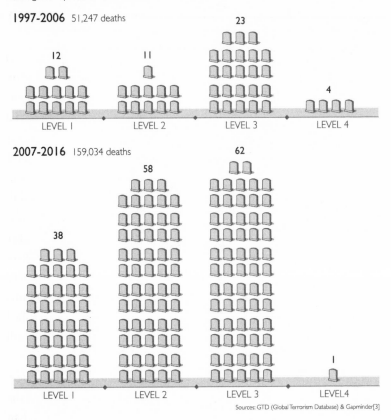

Sources: GTD (Global Terrorism Database) & Gapminder[3]

However, while terrorism has been increasing worldwide, it has actually been decreasing on Level 4. In 2007 to 2016 a total of 1,439
people were killed by terrorists in countries on Level 4. During the ten
years before that, 4,358 were killed. That includes the largest attack
ever, the 2,996 people who died on 9/11 in 2001. Even if we exclude
them, the death toll on Level 4 has remained the same between the

two latest ten-year periods. It was on Levels 1, 2, and 3 that there was a terrible increase in terror-related deaths. Most of that increase was in five countries: Iraq (which accounted for almost half the increase), Afghanistan, Nigeria, Pakistan, and Syria.

Terrorism deaths in the richest countries—i.e., countries on Level 4—accounted for 0.9 percent of all terrorism deaths in 2007 to 2016. They have been decreasing through this century. Since 2001, no terrorist has managed to kill a single individual by hijacking a commercial airline. In fact, it is hard to think of a cause of death that kills fewer people in countries on Level 4 than terrorism. On US soil, 3,172 people died from terrorism over the last 20 years—an average of 159 a year. During those same years, alcohol contributed to the death of 1.4 million people in the United States—an average of 69,000 a year. This is not quite a fair comparison, because in most of those cases the drinker is also the victim. It would be fairer to look only at those deaths where the victim was not the drinker: car accidents and homicide. A very conservative estimate would give us a US figure of roughly 7,500 deaths a year. In the United States, the risk that your loved one will be killed by a drunk person is nearly 50 times higher than the risk he or she will be killed by a terrorist.

But dramatic terrorist incidents in countries on Level 4 receive widespread media coverage that is denied to most victims of alcohol. And the very visible security controls at airports, which make the risk lower than ever, might give an impression of increased danger.

One week after September 11, 2001, according to Gallup, 51 percent of the US public felt worried that a family member would become a victim of terrorism. Fourteen years later, the figure was the same: 51 percent. People are almost as scared today as they were the week after the Twin Towers came down.

Fear vs. Danger: Being Afraid of the Right Things

Fear can be useful, but only if it is directed at the right things. The fear instinct is a terrible guide for understanding the world. It makes us give our attention to the unlikely dangers that we are most afraid of, and neglect what is actually most risky.

This chapter has touched on terrifying events: natural disasters (0.1 percent of all deaths), plane crashes (0.001 percent), murders (0.7 percent), nuclear leaks (0 percent), and terrorism (0.05 percent). None of them kills more than 1 percent of the people who die each year, and still they get enormous media attention. We should of course work to reduce these death rates as well. Still, this helps to show just how much the fear instinct distorts our focus. To understand what we should truly be scared of, and how to truly protect our loved ones from danger, we should suppress our fear instinct and measure the actual death tolls.

Because "frightening" and "dangerous" are two different things. Something frightening poses a perceived risk. Something dangerous poses a real risk. Paying too much attention to what is frightening rather than what is dangerous—that is, paying too much attention to fear—creates a tragic drainage of energy in the wrong directions. It makes a terrified junior doctor think about nuclear war when he should be treating hypothermia, and it makes whole populations focus on earthquakes and crashing planes and invisible substances when millions are dying from diarrhea and seafloors are becoming underwater deserts. I would like my fear to be focused on the mega dangers of today, and not the dangers from our evolutionary past.

Factfulness

Factfulness is . . . recognizing when frightening things get our attention, and remembering that these are not necessarily the most risky. Our natural fears of violence, captivity, and contamination make us systematically overestimate these risks.

To control the fear instinct, **calculate the risks.**

- **The scary world: fear vs. reality.** The world seems scarier than it is because what you hear about it has been selected— by your own attention filter or by the media—precisely because it is scary.

- **Risk = danger × exposure.** The risk something poses to you depends not on how scared it makes you feel, but on a combination of two things. How dangerous is it? And how much are you exposed to it?

- **Get calm before you carry on.** When you are afraid, you see the world differently. Make as few decisions as possible until the panic has subsided.

CHAPTER FIVE

THE SIZE INSTINCT

Putting war memorials and bear attacks in proportion using
two magic tools that you already possess

The Deaths I Do Not See

When I was a young doctor in Mozambique in the early 1980s, I had
to do some very difficult math. The math was difficult because of what
I was counting. I was counting dead children. Specifically, I was comparing the number of deaths among children admitted to our hospital in Nacala with the number of children dying in their homes within
the district we were supposed to serve.

At that time, Mozambique was the poorest country in the world.
In my first year in Nacala district, I was the only doctor for a population of 300,000 people. In my second year, a second doctor joined
me. We covered a population that in Sweden would have been served

by 100 doctors, and every morning on my way to work I said to myself, "Today I must do the work of 50 doctors."

We admitted around 1,000 very sick children each year to the district's one small hospital, which meant around three per day. I will never forget trying to save the lives of those children. All had very severe diseases like diarrhea, pneumonia, and malaria, often complicated by anemia and malnutrition, and despite our best efforts, around one in 20 of them died. That was one child every week, almost all of whom we could have cured if we had had more and better resources and staff.

The care we could provide was rudimentary: water and salt solutions and intramuscular injections. We did not give intravenous drips: the nurses had not yet acquired the skills to administer them and it would have taken up too much of the doctors' time to place and supervise the infusions. We rarely had oxygen tanks and we had limited capacity for blood transfusions. This was the medicine of extreme poverty.

One weekend, a friend came to stay with us—a Swedish pediatrician who worked in the slightly better hospital in a bigger city 200 miles away. On the Saturday afternoon, I had to go on an emergency call to the hospital and he came with me. When we arrived, we were met by a mother with fear in her eyes. In her arms was her baby who had severe diarrhea and was so weak that she could not breastfeed. I admitted the child, inserted a feeding tube, and ordered that oral rehydration solution should be given through the tube. My pediatrician friend dragged me into the corridor by the arm. He was very upset and angrily challenged the substandard treatment I had prescribed, accusing me of skimping in order to get home for dinner. He wanted me to give the baby an intravenous drip.

I became angry at his lack of understanding. "This is our standard treatment here," I explained. "It would take me half an hour to get a drip running for this child and then there would be a high risk that the nurse would mess it up. And yes, I do have to get home for dinner sometimes, otherwise my family and I would not last here more than a month."

My friend couldn't accept it. He decided to stay at the hospital struggling for hours to get a needle into a tiny vein.

When my colleague finally joined me back at home, the debate continued. "You must do everything you can for every patient who presents at the hospital," he urged.

"No," I said. "It is unethical to spend all my time and resources trying to save those who come here. I can save more children if I improve the services outside the hospital. I am responsible for *all* the child deaths in this district: the deaths I do not see just as much as the deaths in front of my eyes."

My friend disagreed, as do most doctors and perhaps most members of the public. "Your obligation is to do everything for the patients in your care. Your claim that you can save more children elsewhere is just a cruel theoretical guess." I was very tired. I stopped arguing and went to bed, but the next day I started counting.

Together with my wife, Agneta, who managed the delivery ward, I did the math. We knew that a total of 946 children had been admitted to the hospital that year, almost all of them below the age of five, and of those, 52 (5 percent) had died. We needed to compare that number with the number of child deaths in the whole district.

The child mortality rate of Mozambique was then 26 percent. There was nothing special about Nacala district, so we could use that figure. The child mortality rate is calculated by taking the number of child deaths in a year and dividing it by the number of births in that year.

So if we knew the number of births in the district that year, we could estimate the number of child deaths, using the child mortality rate of 26 percent. The latest census gave us a number for births in the city: roughly 3,000 each year. The population of the district was five times the population of the city, so we estimated there had probably been five times as many births: 15,000. So 26 percent of that number told us that I was responsible for trying to prevent 3,900 child deaths every year, of which 52 happened in the hospital. I was seeing only 1.3 percent of my job.

Now I had a number that supported my gut feeling. Organizing, supporting, and supervising basic community-based health care that could treat diarrhea, pneumonia, and malaria before they became life-threatening would save many more lives than putting drips on terminally ill children in the hospital. It would, I believed, be truly unethical to spend more resources in the hospital before the majority of the population—and the 98.7 percent of dying children who never reached the hospital—had some form of basic health care.

So we worked to train village health workers, to get as many children as possible vaccinated, and to treat the main child killers as early as possible in small health facilities that could be reached even by mothers who had to walk.

This is the cruel calculus of extreme poverty. It felt almost inhuman to look away from an individual dying child in front of me and toward hundreds of anonymous dying children I could not see.

I remember the words of Ingegerd Rooth, who had been working as a missionary nurse in Congo and Tanzania before she became my mentor. She always told me, "In the deepest poverty you should never do anything perfectly. If you do you are stealing resources from where they can be better used."

Paying too much attention to the individual visible victim rather than to the numbers can lead us to spend all our resources on a fraction of the problem, and therefore save many fewer lives. This principle applies anywhere we are prioritizing scarce resources. It is hard for people to talk about resources when it comes to saving lives, or prolonging or improving them. Doing so is often taken for heartlessness. Yet so long as resources are not infinite—and they never are infinite—it is the most compassionate thing to do to use your brain and work out how to do the most good with what you have.

This chapter is full of data about dead children because saving children's lives is what I care about most in the whole world. It seems heartless and cruel, I know, to count dead children and to talk about cost-effectiveness in the same sentence as a dying child. But if you think about it, working out the most cost-effective way of saving as many children's lives as possible is the least heartless exercise of them all.

Just as I have urged you to look behind the statistics at the individual stories, I also urge you to look behind the individual stories at the statistics. The world cannot be understood without numbers. And it cannot be understood with numbers alone.

The Size Instinct

You tend to get things out of proportion. I do not mean to sound rude. Getting things out of proportion, or misjudging the size of things, is something that we humans do naturally. It is instinctive to look at a lonely number and misjudge its importance. It is also instinctive—like in the hospital in Nacala—to misjudge the importance of a single instance or an identifiable victim. These two tendencies are the two key aspects of the size instinct.

The media is this instinct's friend. It is pretty much a journalist's professional duty to make any given event, fact, or number sound more important than it is. And journalists know that it feels almost inhuman to look away from an individual in pain.

The two aspects of the size instinct, together with the negativity instinct, make us systematically underestimate the progress that has been made in the world. In the test questions about global proportions, people consistently say about 20 percent of people are having their basic needs met. The correct answer in most cases is close to 80 percent or even 90 percent. Proportion of children vaccinated: 88 percent. Proportion of people with electricity: 85 percent. Proportion of girls in primary school: 90 percent. The use of numbers that sound enormous, together with constant images of individual suffering presented by the charities and the media, distort people's view of the world and they systematically underestimate all these proportions and all this progress.

At the same time, we systematically overestimate other proportions. The proportion of immigrants in our countries. The proportion of people opposed to homosexuality. In each of these cases, at least in the United States and Europe, our interpretations are more dramatic than the reality.

The size instinct directs our limited attention and resources toward those individual instances or identifiable victims, those concrete things right in front of our eyes. Today there are robust data sets for making the kinds of comparisons I made in Nacala on a global scale, and they show the same thing: It is not doctors and hospital beds that save children's lives in countries on Levels 1 and 2. Beds and doctors are easy to count and politicians love to inaugurate buildings. But almost all the increased child survival is achieved through preventive measures outside hospitals by local nurses, midwives, and well-educated parents. Especially mothers: the data shows that half the increase in child survival in the world happens because the mothers can read and write. More children now survive because they don't get ill in the first place. Trained midwives assist their mothers during pregnancy and delivery. Nurses immunize them. They have enough food, their parents keep them warm and clean, people around them wash their hands, and their mothers can read the instructions on that jar of pills. So if you are investing money to improve health on Level 1 or 2, you should put it into primary schools, nurse education, and vaccinations. Big impressive-looking hospitals can wait.

How to Control the Size Instinct

To avoid getting things out of proportion you need only two magic tools: comparing and dividing. What did you say? You already know both of them? Great, then all you need is to start using them. Make it a habit! I'll show you how.

Compare the Numbers

The most important thing you can do to avoid misjudging something's importance is to avoid lonely numbers. Never, ever leave a number all by itself. Never believe that one number on its own can be meaningful. If you are offered one number, always ask for at least one more. Something to compare it with.

Be especially careful about big numbers. It is a strange thing, but numbers over a certain size, when they are not compared with anything else, always look big. And how can something big not be important?

4.2 Million Dead Babies

Last year, 4.2 million babies died.

That is the most recent number reported by UNICEF of deaths before the age of one, worldwide. We often see lonely and emotionally charged numbers like this in the news or in the materials of activist groups or organizations. They produce a reaction.

Who can even imagine 4.2 million dead babies? It is so terrible, and even worse when we know that almost all died from easily preventable diseases. And how can anyone argue that 4.2 million is anything other than a huge number? You might think that nobody would even try to argue that, but you would be wrong. That is exactly why I mentioned this number. Because it is *not* huge: it is beautifully small.

If we even start to think about how tragic each of these deaths is for the parents who had waited for their newborn to smile, and walk, and play, and instead had to bury their baby, then this number could keep us crying for a long time. But who would be helped by these tears? Instead let's think clearly about human suffering.

The number 4.2 million is for 2016. The year before, the number was 4.4 million. The year before that, it was 4.5 million. Back in 1950, it was 14.4 million. That's almost 10 million more dead babies per year, compared with today. Suddenly this terrible number starts to look smaller. In fact the number has never been lower since the measuring began.

Of course, I am the first person to wish the number was even lower and falling even faster. But to know how to act, and how to prioritize resources, nothing can be more important than doing the cool-headed math and realizing what works and what doesn't. And this is clear: more and more deaths are being prevented. We would never realize that without comparing the numbers.

A Large War

The Vietnam War was the Syrian war of my generation.

Two days before Christmas in 1972, seven bombs killed 27 patients and members of staff at the Bach Mai hospital in Hanoi in Vietnam. I was studying medicine in Uppsala in Sweden. We had plenty of medical equipment and yellow blankets. Agneta and I coordinated a collection, which we packed in boxes and sent to Bach Mai.

Fifteen years later, I was in Vietnam to evaluate a Swedish aid project. One lunchtime, I was eating my rice next to one of my local colleagues, a doctor named Niem, and I asked him about his background. He told me he had been inside the Bach Mai hospital when the bombs fell. Afterward, he had coordinated the unpacking of boxes of supplies that had arrived from all over the world. I asked him if he remembered some yellow blankets and I got goose bumps as he described the fabric's pattern to me. It felt like we had been friends forever.

At the weekend, I asked Niem to show me the monument to the Vietnam War. "You mean the 'Resistance War Against America,'" he said. Of course, I should have realized he wouldn't call it the Vietnam War. Niem drove me to one of the city's central parks and showed me a small stone with a brass plate, three feet high. I thought it was a joke. The protests against the Vietnam War had united a generation of activists in the West. It had moved me to send blankets and medical equipment. More than 1.5 million Vietnamese and 58,000 Americans had died. Was this how the city commemorated such a catastrophe? Seeing that I was disappointed, Niem drove me to see a bigger monument: a marble stone, 12 feet high, to commemorate independence from French colonial rule. I was still underwhelmed.

Then Niem asked me if I was ready to see the proper war monument. He drove a little way further, and pointed out of the window. Above the treetops I could see a large pagoda, covered in gold. It seemed about 300 feet high. He said, "Here is where we commemorate our war heroes. Isn't it beautiful?" This was the monument to Vietnam's wars with China.

The wars with China had lasted, on and off, for 2,000 years. The French occupation had lasted 200 years. The "Resistance War Against America" took only 20 years. The sizes of the monuments put things in perfect proportion. It was only by comparing them that I could understand the relative insignificance of "the Vietnam War" to the people who now live in Vietnam.

Bears and Axes

Mari Larsson was 38 years old when she was killed by multiple blows to the head from an axe. It was the night of October 17, 2004. Mari's former partner had broken into her house in the small town of Piteå in the north of Sweden and was waiting for her to come home. The tragic and brutal murder of a mother of three was barely reported in the national media and even the local newspaper gave it only modest coverage.

That same day a 40-year-old father of three, also living in the far north of Sweden, was killed by a bear while out hunting. His name was Johan Vesterlund and he was the first person killed by a bear in Sweden since 1902. This brutal, tragic, and, crucially, rare event received massive coverage throughout Sweden.

In Sweden, a fatal bear attack is a once-in-a-century event. Meanwhile, a woman is killed by her partner every 30 days. This is a 1,300-fold difference in magnitude. And yet one more domestic murder had barely registered, while the hunting death was big news.

Despite what the media coverage might make us think, each death was equally tragic and horrendous. Despite what the media might make us think, people who care about saving lives should be much more concerned about domestic violence than about bears.

It seems obvious when you compare the numbers.

Tuberculosis and Swine Flu

It is not only bears and axes that the news media gets out of proportion.

In 1918 the Spanish flu killed around 2.7 percent of the world population. The risk of an outbreak of a flu against which we have no vaccine remains a constant threat, which we should all take extremely seriously. In the first months of 2009, thousands of people died from the swine flu. For two weeks it was all over the news. Yet, unlike with Ebola in 2014, the number of cases did not double. It did not even go up in a straight line. I and others concluded this flu was not as aggressive as the first alarm had indicated. But journalists kept the fear boiling for several weeks.

Finally I got tired of the hysteria and calculated the rate of news reports versus fatalities. Over a period of two weeks, 31 people had died from swine flu, and a news search on Google brought up 253,442 articles about it. That was 8,176 articles per death. Over the same two-week period, I calculated that roughly 63,066 people had died of tuberculosis (TB). Almost all these people were on Levels 1 and 2, where TB remains a major killer even though it can now be treated. But TB is infectious and TB strains can become resistant and kill many people on Level 4. The news coverage for TB was at a rate of 0.1 article per death. Each swine flu death received 82,000 times more attention than each equally tragic death from TB.

The 80/20 Rule

It's so easy to get things out of proportion, but luckily there are also some easy solutions. Whenever I have to compare lots of numbers and work out which are the most important, I use the simplest-ever thinking tool. I look for the largest numbers.

That is all there is to the 80/20 rule. We tend to assume that all items on a list are equally important, but usually just a few of them are more important than all the others put together. Whether it is causes of death or items in a budget, I simply focus first on understanding those that make up 80 percent of the total. Before I spend time on the smaller ones, I ask myself: Where are the 80 percent? Why are these so big? What are the implications?

For example, here's a list of the world's energy sources, in alphabetical order: biofuels, coal, gas, geothermal, hydro, nuclear, oil, solar, wind. Presented like that, they all seem equally important. If we instead sort them according to how many units of energy they generate for humanity, three outnumber all the rest, as this graph shows.

GLOBAL ENERGY SOURCES, 2016

Energy consumption worldwide in 100 terawatt-hours (TWh)

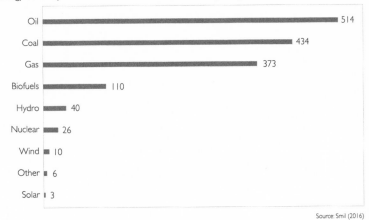

Oil 514
Coal 434
Gas 373
Biofuels 110
Hydro 40
Nuclear 26
Wind 10
Other 6
Solar 3

Source: Smil (2016)

To give myself the big picture I would use the 80/20 rule, which tells us that oil+coal+gas give us more than 80 percent of our energy: 87 percent in fact.

I first discovered how useful the 80/20 rule is when I started to review aid projects for the Swedish government. In most budgets, around 20 percent of the lines sum up to more than 80 percent of the total. You can save a lot of money by making sure you understand these lines first.

Doing just that is how I discovered that half the aid budget of a small health center in rural Vietnam was about to be spent on 2,000 of the wrong kind of surgical knives. It's how I discovered that 100 times too much—4 million liters—of baby formula was about to be sent to a refugee camp in Algeria. And it is how I stopped 20,000 testicular prostheses from being sent to a small youth clinic in Nicaragua. In each case I simply looked for the biggest single items taking up 80 percent of the budget, then dug down into any that seemed unusual. In each case the problem was due to a simple confusion or tiny error such as a missing decimal point.

The 80/20 rule is as easy as it seems. You just have to remember to use it. Here's one more example.

The PIN Code of the World

We can understand the world better, and make better decisions about it, if we know where the biggest proportion of the population lives now and where it will live in the future. Where is the world market? Where are the internet users? Where will tourists come from in the future? Where are most of the cargo ships going? And so on.

FACT QUESTION 8

There are roughly 7 billion people in the world today. Which map shows best where they live? (Each figure represents 1 billion people.)

WHERE PEOPLE LIVE The world population in billions of people

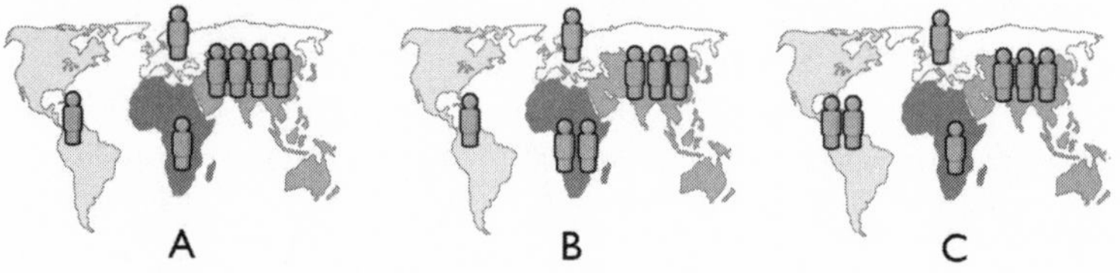

A B C

This is one of the fact questions where people score best. They are almost as good as the chimps. Their answers are almost as good as random. By this point in the book, that looks like a great achievement. You see, it all depends on how you compare!

Seventy percent of people still pick the wrong maps, showing 1 billion people on the wrong continent. Seventy percent of people don't know that the majority of mankind lives in Asia. If you really care about a sustainable future or the plundering of our planet's natural resources or the global market, how can you afford to lose track of a billion people?

The correct map is A. The PIN code of the world is 1-1-1-4. That's how to remember the map. From left to right, the number of billions, as a PIN code. Americas: 1, Europe: 1, Africa: 1, Asia: 4. (I have rounded the numbers.) Like all PIN codes, this one will change. By the end of this century, the UN expects there to have been almost no change in the Americas and Europe but 3 billion more people in Africa and 1 billion more in Asia. By 2100 the new PIN code of the world will be 1-1-4-5. More than 80 percent of the world's population will live in Africa and Asia.

If the UN forecasts for population growth are correct, and if incomes in Asia and Africa keep growing as now, then the center of gravity of the world market will shift over the next 20 years from the Atlantic to the Indian Ocean. Today, the people living in rich countries around the North Atlantic, who represent 11 percent of the world population, make up 60 percent of the Level 4 consumer market. Already by 2027, if incomes keep growing worldwide as they are doing now, then that figure will have shrunk to 50 percent. By 2040, 60 percent of Level 4 consumers will live outside the West. Yes, I think the Western domination of the world economy will soon be over.

People in North America and Europe need to understand that most of the world population lives in Asia. In terms of economic muscles "we" are becoming the 20 percent, not the 80 percent. But many of "us" can't fit these numbers into our nostalgic minds. Not only do we misjudge how big our war monuments should be in Vietnam, we also misjudge our importance in the future global marketplace. Many of us forget to behave properly with those who will control the future trade deals.

SOON, MOST PEOPLE ON LEVEL 4 WILL BE NON-WESTERNERS
The world population divided into West and Rest, distributed over incomes.

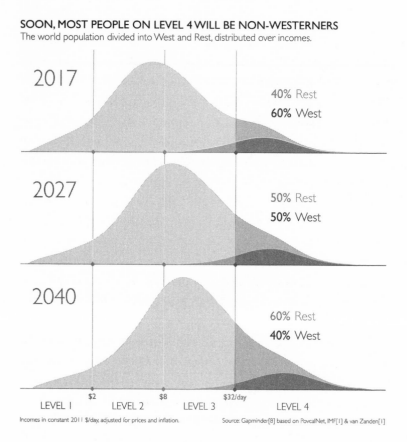

2017 — 40% Rest / 60% West
2027 — 50% Rest / 50% West
2040 — 60% Rest / 40% West

LEVEL 1 — $2 — LEVEL 2 — $8 — LEVEL 3 — $32/day — LEVEL 4

Incomes in constant 2011 $/day, adjusted for prices and inflation.   Source: Gapminder[8] based on PovcalNet, IMF[1] & van Zanden[1]

Divide the Numbers

Often the best thing we can do to make a large number more meaningful is to divide it by a total. In my work, often that total is the total
population. When we divide an amount (say, the number of children
in Hong Kong) by another amount (say, the number of schools in Hong
Kong), we get a rate (children per school in Hong Kong). Amounts
are easier to find because they are easier to produce. Somebody just
needs to count something. But rates are often more meaningful.

The Trend Below the Division Line

I want to return to the 4.2 million dead infants. Earlier in the chapter we compared 4.2 million babies to the 14.4 million who died in 1950. What if fewer children are being born every year and that's the reason fewer babies are dying? When you see one number falling it is sometimes actually because some other background number is falling. To check, we need to divide the total number of child deaths by the total number of births.

In 1950, 97 million children were born and 14.4 million children died. To get the child mortality rate, we divide the number of deaths (14.4 million) by the number of births (97 million). That comes out to 15 percent. So in 1950, out of every 100 babies who were born, 15 died before their first birthday.

Now let's look at the most recent numbers. In 2016, 141 million children were born and 4.2 million died. Dividing the number of births by the number of deaths comes out to just 3 percent. Out of every 100 babies born across the world, only three die before reaching the age of one. Wow! The infant mortality rate has changed from 15 percent to 3 percent. When we compare rates, rather than amounts of dead children, the most recent number suddenly seems astonishingly low.

Some people feel ashamed when doing this kind of math with human lives. I feel ashamed when not doing it. A lonely number always makes me suspicious that I will misinterpret it. A number that I have compared and divided can instead fill me with hope.

Per Person

"The forecasts show that it is China, India, and the other emerging economies that are increasing their carbon dioxide emissions at a speed that will cause dangerous climate change. In fact, China already emits more CO_2 than the USA, and India already emits more than Germany."

This outspoken statement came from an environment minister from a European Union country who was part of a panel discussing climate change at the World Economic Forum in Davos in January 2007. He made his attribution of blame in a neutral tone of voice, as if he were stating a self-evident fact. Had he been watching the faces of the Chinese and Indian panel members he would have realized that his view was not self-evident at all. The Chinese expert looked angry but continued to stare straight ahead. The Indian expert, in contrast, could not sit still. He waved his arm and could barely wait for the moderator's signal that he could speak.

He stood up. There was a short silence while he looked into the face of each panel member. His elegant dark blue turban and expensive-looking dark gray suit, and the way he was behaving in his moment of outrage, confirmed his status as one of India's highest-ranking civil servants with many years' experience as a lead expert at the World Bank and the International Monetary Fund. He made a sweeping gesture toward the panel members from the rich nations and then said loudly and accusingly, "It was you, the richest nations, that put us all in this delicate situation. You have been burning increasing amounts of coal and oil for more than a century. You and only you pushed us to the brink of climate change." Then he suddenly changed posture, put his palms together in an Indian greeting, bowed, and almost whispered in a very kind voice, "But we forgive you, because you did not know what you were doing. We should never blame someone retrospectively for harm they were unaware of." Then he straightened up and delivered his final remark as a judge giving his verdict, emphasizing each word by slowly moving his raised index finger. "But from now on we count carbon dioxide emission *per person.*"

I couldn't have agreed more. I had for some time been appalled by the systematic blaming of climate change on China and India based on *total* emissions per nation. It was like claiming that obesity was worse in China than in the United States because the total bodyweight of the Chinese population was higher than that of the US population. Arguing about emissions per nation was pointless when there was such enormous variation in population size. By this logic, Norway, with its population of 5 million, could be emitting almost any amount of carbon dioxide per person.

In this case, the large numbers—total emissions per nation—needed to be divided by the population of each country to give meaningful and comparable measures. Whether measuring HIV, GDP, mobile phone sales, internet users, or CO_2 emissions, a per capita measurement—i.e., a rate per person—will almost always be more meaningful.

It's Dangerous Out There

The safest lives in history are lived today by people on Level 4. Most preventable risks have been eliminated. Still, many walk around feeling worried.

They worry about all kinds of dangers "out there." Natural disasters kill so many people, diseases spread, and airplanes crash. They all happen all the time out there, beyond the horizon. It's a bit strange, isn't it? Such terrifying things rarely happen "here," in this safe place where we live. But out there, they seem to happen every day. Remember, though, "out there" is the sum of millions of places, while you live in just one place. Of course more bad things happen out there: out there is much bigger than here. So even if all the places out there were just as safe as your place, hundreds of terrible events would still happen there. If you could keep track of each separate place though, you would be surprised how peaceful most of them were. Each of them shows up on your screen only on that single day when something terrible happens. All the other days, you don't hear about them.

Compare and Divide

When I see a lonely number in a news report, it always triggers an alarm: What should this lonely number be compared to? What was that number a year ago? Ten years ago? What is it in a comparable country or region? And what should it be divided by? What is the total of which this is a part? What would this be per person? I compare the rates, and only then do I decide whether it really is an important number.

Factfulness

Factfulness is . . . recognizing when a lonely number seems impressive (small or large), and remembering that you could get the opposite impression if it were compared with or divided by some other relevant number.

To control the size instinct, **get things in proportion.**

- **Compare.** Big numbers always look big. Single numbers on their own are misleading and should make you suspicious. Always look for comparisons. Ideally, divide by something.
- **80/20.** Have you been given a long list? Look for the few largest items and deal with those first. They are quite likely more important than all the others put together.
- **Divide.** Amounts and rates can tell very different stories. Rates are more meaningful, especially when comparing between different-sized groups. In particular, look for rates per person when comparing between countries or regions.

CHAPTER SIX

THE GENERALIZATION INSTINCT

OQQ

Why I had to lie about the Danes, and how it
can be smart to build half a house

Dinner Is Served

An orange sun was setting behind the acacia trees on the savanna of
the Bandundu region south of the Congo River, half a day's walk from
the end of the paved road. This is where you find the people who live in
extreme poverty: they are stuck behind that mountain, beyond where
the road ends. My colleague Thorkild and I had spent the day interviewing the people in this remote village about their nutrition, and
now they wanted to throw us a party. No one had ever walked so far to
ask them about their problems.

As Swedish villagers would have done 100 years ago, they were
demonstrating their gratitude and respect by serving their guests the

biggest piece of meat they could find. The entire village was gathered in a circle around Thorkild and me as we were presented with our plates. On top of two large green leaves lay two whole, skinned, grilled rats.

I thought I might throw up. Then I noticed that Thorkild had already started eating: we were both very hungry after a whole day's work with no food. I looked around at the villagers who were smiling at me expectantly. I had to eat it, and I did. It was actually not that bad: it tasted a bit like chicken. To be polite, I tried to look happy as I swallowed it down.

Then it was time for dessert: another plate, full of big, white larvas from the palm nut tree. And I do mean big—each one was longer and thicker than my thumb, and had been lightly fried in its own fat. But I wondered, had they been *too* lightly fried? Because they seemed to be moving. The villagers were proud to offer us such a delicious treat.

Remember, I am a sword swallower. I should be able to push anything down my throat. And I am not usually a fussy eater: I had even once eaten porridge made from mosquitos. But no. This, I couldn't do. The heads of the larvas looked like little brown nuts and their thick bodies like transparent wrinkled marshmallows, through which I could see their intestines. The villagers gestured that I should bite them in two and suck out the insides. If I tried I would puke the rat back up. I did not want to offend.

Suddenly, an idea. I smiled softly and said regretfully, "You know what, I am sorry, but I can't eat larvas."

Thorkild turned to me, surprised. He already had a couple of larvas hanging out of the corners of his mouth. He really loved those larvas. He had previously worked as a missionary in Congo, where they had been the highlight of every week for one whole year.

"You see, we don't eat larvas," I said, trying to look convincing. The villagers looked at Thorkild.

"But he eats them?" they asked. Thorkild stared at me.

"Ah," I said. "You see, he comes from a different tribe. I come from Sweden, he comes from Denmark. In Denmark, they love eating larvas. But in Sweden it's against our culture." The village teacher went and got out the world map and I pointed out the water separating our two countries. "On this side of the water they eat larvas," I said, "and on this side we don't." It's actually one of the most blatant lies I have ever told, but it worked. The villagers were happy to share my dessert between them. Everyone, everywhere knows that people from different tribes have different customs.

The Generalization Instinct

Everyone automatically categorizes and generalizes all the time. Unconsciously. It is not a question of being prejudiced or enlightened. Categories are absolutely necessary for us to function. They give structure to our thoughts. Imagine if we saw every item and every scenario as truly unique—we would not even have a language to describe the world around us.

The necessary and useful instinct to generalize, like all the other instincts in this book, can also distort our worldview. It can make us mistakenly group together things, or people, or countries that are actually very different. It can make us assume everything or everyone in one category is similar. And, maybe most unfortunate of all, it can make us jump to conclusions about a whole category based on a few, or even just one, unusual example.

Once again, the media is the instinct's friend. Misleading generalizations and stereotypes act as a kind of shorthand for the media, providing quick and easy ways to communicate. Here are just a few examples from today's newspaper: rural life, middle class, super mom, gang member.

When many people become aware of a problematic generalization it is called a stereotype. Most commonly, people talk about race and gender stereotyping. These cause many very important problems, but they are not the only problems caused by wrong generalizations. Wrong generalizations are mind-blockers for all kinds of understanding.

The gap instinct divides the world into "us" and "them," and the generalization instinct makes "us" think of "them" as all the same.

Are you working for a commercial company on Level 4? There's a great risk you're missing the majority of your potential consumers and producers because of your generalizations. Are you working in finance in a big bank? There's a great risk you are investing your clients' money in the wrong places, because you're bundling together people who are vastly different.

FACT QUESTION 9

How many of the world's 1-year-old children today have been vaccinated against some disease?

- ☐ A: 20 percent
- ☐ B: 50 percent
- ☐ C: 80 percent

To compare ignorance between different kinds of experts, the regular polling companies couldn't help me. They don't have access to the staff of big corporations and government organizations. That's one reason I started polling my audience at the start of my lectures. I have tested a total of 12,596 people at 108 lectures over the last five years. This question gets the worst results. Look at the table on the next page, where I have ranked 12 groups of experts according to how many picked the most incorrect answer.

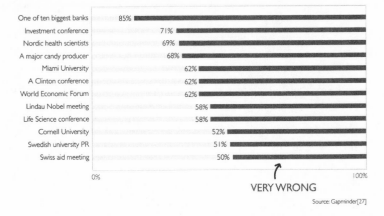

FACT QUESTION 2 RESULTS: percentage who answered very wrong.
How many of the world's 1-year-old children today have been vaccinated against some disease?
(Correct answer: 80%. Very wrong answer: 20%.)

Source: Gapminder[27]

The worst results come from an annual gathering of global finance managers at the headquarters of one of the world's ten largest banks. I have visited three of them. I can't tell you which one this was, because I signed a piece of paper. A roaring 85 percent of the 71 well-dressed bankers in the room believed that a minority of the world's children had been vaccinated. An extremely wrong answer.

Vaccines must be kept cold all the way from the factory to the arm of the child. They are shipped in refrigerated containers to harbors around the world, where they get loaded into refrigerated trucks. These trucks take them to local health clinics, where they are stored in refrigerators. These logistic distribution paths are called cool chains. For cool chains to work, you need all the basic infrastructure for transport, electricity, education, and health care to be in place. This is exactly the same infrastructure needed to establish new factories. The fact that 88 percent are vaccinated but major financial investors believe it is only 20 percent indicates that there is a big chance they are failing at their jobs by missing out on huge investment opportunities (probably the most profitable ones in the fastest-growing parts of the world).

You make this kind of false assumption when you have a "them" category in your head, into which you put the majority of humanity. What images are you using to imagine what life is like in this category? Are you perhaps recalling the most vivid and disturbing images from the news? I think that is exactly what's going on when people on Level 4 answer this badly on this kind of fact question. The extreme deprivation we see on the news ends up stereotyping the majority of mankind.

Every pregnancy results in roughly two years of lost menstruation. If you are a manufacturer of menstrual pads, this is bad for business. So you ought to know about, and be so happy about, the drop in babies per woman across the world. You ought to know and be happy too about the growth in the number of educated women working away from home. Because these developments have created an exploding market for your products over the last few decades among billions of menstruating women now living on Levels 2 and 3.

But, as I realized when I attended an internal meeting at one of the world's biggest manufacturers of sanitary wear, most Western manufacturers have completely missed this. Instead, when hunting for new customers they are often stuck dreaming up new needs among the 300 million menstruating women on Level 4. "What if we market an even thinner pad for bikinis? What about pads that are invisible, to wear under Lycra? How about one pad for each kind of outfit, each situation, each sport? Special pads for mountain climbers!" Ideally, all the pads are so small they need to be replaced several times a day. But like most rich consumer markets, the basic needs are already met, and producers fight in vain to create demand in ever-smaller segments.

Meanwhile, on Levels 2 and 3, roughly 2 billion menstruating women have few alternatives to choose from. These women don't wear Lycra and won't spend money on ultrathin pads. They demand a low-cost pad that will be reliable throughout the day so they don't have to change it when they are out at work. And when they find a product they like, they will probably stick to that brand for their whole lives and recommend it to their daughters.

The same logic applies to many other consumer products, and I have given hundreds of lectures to business leaders making this same point. The majority of the world population is steadily moving up the levels. The number of people on Level 3 will increase from two billion to four billion between now and 2040. Almost everyone in the world is becoming a consumer. If you suffer from the misconception that most of the world is still too poor to buy anything at all, you risk missing out on the biggest economic opportunity in world history while you use your marketing spend to push special "yoga" pads to wealthy hipsters in the biggest cities in Europe. Strategic business planners need a fact-based worldview to find their future customers.

Reality Bites

You need the generalization instinct to live your everyday life, and occasionally it can save you from having to eat something disgusting. We always need categories. The challenge is to realize which of our simple categories are misleading—like "developed" and "developing" countries—and replace them with better categories, like the four levels.

One of the best ways to do this is to travel, if you possibly can. That's why I made my global health students from Karolinska Institutet, the medical university in Stockholm, go on study visits to countries on Levels 1, 2, and 3, where they attended university courses, visited hospitals, and stayed with local families. Nothing beats a first-hand experience.

Those students are usually privileged young Swedes who want to do good in the world but don't really know the world. Some of them say they have traveled: often they have had a cappuccino at a café next to an eco-tourism agency, but never entered a single family home.

On day one of a trip to Thiruvananthapuram in Kerala in India, or Kampala in Uganda, they usually express surprise that the city is so well organized. There are traffic lights and sewage systems and no one is dying in the street.

On day two, we usually visit a public hospital. When they see that there is no paint on the walls and no air-conditioning and 60 people to a room, my students whisper to each other that this place must be extremely poor. I have to explain that people living in extreme poverty have no hospitals at all. A woman living in extreme poverty gives birth on a mud floor, attended by a midwife with no training who has walked barefoot in the dark. The hospital administrator helps. She explains that not painting the walls can be a strategic decision in countries on Levels 2 and 3. It's not that they can't afford the paint. Flaking walls keep away the richer patients and their time-consuming demands for costly treatments, allowing hospitals to use their limited resources to treat more people in more cost-effective ways.

My students then learn that one of the patients cannot afford to pay for the insulin he has been prescribed for his newly diagnosed diabetes. The students don't understand: this must be an advanced hospital if it can diagnose diabetes. But how bizarre if the patient cannot then afford the treatment. Yet this is very common on Level 2: the public health system can pay for some diagnosis, for emergency care, and for inexpensive drugs. This leads to great improvements in survival rates. But there's simply not enough money (unless the costs come down) for expensive treatments for lifelong conditions like diabetes.

On one particular occasion a student's misunderstanding of life in countries on Level 2 nearly cost her very dearly. We were visiting a beautiful and modern private hospital in Kerala, India, eight stories tall. We waited some time in the lobby for a student in our group who was late. After 15 minutes, we decided not to wait for her any longer and walked down a corridor and got into a large elevator, big enough to take several hospital beds. Our host, the head of the intensive care unit, pressed the button for the sixth floor. Just as the doors were sliding closed, we saw the young blond Swede rush into the hospital lobby.

"Come, run faster!" shouted her friend from the door of the elevator, and she stretched her leg out to stop the doors from closing. Everything then happened very quickly. The doors just continued to close tightly around my student's leg. She cried out in pain and fear. The elevator started moving upward. She cried out louder. Just as I realized this young woman's leg was going to get crushed against the top of the doorway, our host leaped across the elevator and hit the red emergency stop button. He hissed at me to help and between us we prised the doors far enough apart to release my student's bleeding limb.

Afterward, our host looked at me and said, "I have never seen that before. How can you admit such stupid people for medical training?" I explained that all elevators in Sweden had sensors on the doors. If something were put between them, they would instantaneously stop closing and open instead. The Indian doctor looked doubtful. "But how can you be sure that this advanced mechanism is working every single time?" I felt stupid with my reply: "It just always does. I suppose it's because there are strict safety rules and regular inspections." He didn't look convinced. "Hmmm. So your country has become so safe that when you go abroad the world is dangerous for you."

I can assure you that the young woman was not at all stupid. She had simply, and unwisely, generalized from her own Level 4 experience of elevators to all elevators in all countries.

On the last day, we have a little ceremony to say goodbye where I sometimes learn something about the generalizations other people make about us. On this particular occasion in India, my female students arrived on time, beautifully dressed in colorful saris they had bought locally. (The elevator-door leg injury was nicely healed.) They were followed ten minutes later by the male students, evidently hungover and dressed in torn jeans and dirty T-shirts. India's leading professor of forensic medicine leaned over to me and whispered, "I hear you have love marriages in your country but that must be a lie. Look at these men. What woman would marry them if their parents didn't make them?"

When visiting reality in other countries, and not just the backpacker cafés, you realize that generalizing from what is normal in your home environment can be useless or even dangerous.

My First Time

I do not mean to sound critical about my students. I am no better myself.

In 1972, as a fourth-year medical student, I studied at the medical school in Bangalore. The first class I attended was on examining kidney X-rays. Looking at the first image, I realized this must be kidney cancer. I decided to wait awhile before telling the class, out of respect. I didn't want to show off. Several hands then went into the air and the Indian students one by one explained how best to diagnose this cancer, how and where it usually spreads, and how best to treat it. On and on they went for 30 minutes, answering questions I thought only chief physicians knew. I realized my embarrassing mistake. I must have come to the wrong room. These must not be fourth-year students, these must be specialists. I had nothing to add to their analysis.

On our way out, I told a fellow student I was supposed to be with the fourth-years. "That's us," he said. I was stunned. They had caste marks on their foreheads and lived where exotic palm trees grew. How could they know much more than me? Over the next few days I learned that they had a textbook three times as thick as mine, and they had read it three times as many times.

I remember this whole experience as the first time in my life that I suddenly had to change my worldview: my assumption that I was superior because of where I came from, the idea that the West was the best and the rest would never catch up. At that moment, 45 years ago, I understood that the West would not dominate the world for much longer.

How to Control the Generalization Instinct

If you can't travel, please do not worry. There are other ways to avoid using wrong categories.

Find Better Categories: Dollar Street

Anna would always insist that the trips I did with my students were a naïve and unrealistic way to teach most people about the world. Few people wanted to spend their hard-earned money traveling to far-flung places only to try a pit latrine and experience the unglamorous everyday life on Levels 1, 2, or 3, far from the beach, the great cuisine and bars, and the fairy-tale-like wildlife.

Most people were just as uninterested in studying the data about global trends and proportions. And anyway, even looking at the data, it was pretty hard to understand what it meant for everyday life on different levels.

Remember the photos used to describe the levels in the chapter on the gap instinct? They all come from Dollar Street, a project that Anna developed to teach armchair travelers about the world. Now you can understand how people live without leaving your home.

Imagine all the homes in the world lined up on one long street, sorted by income. The poorest live at the left end of the street and the richest live at the right end. Everybody else? Of course, you know it by now: most people live somewhere in the middle. Your house number on this street represents your income. Your neighbors on Dollar Street are people from all over the world with the same income as you.

Anna has so far sent photographers out to visit about 300 families in more than 50 countries. Their photos document how people eat, sleep, brush their teeth, and prepare food. They capture what their homes are made of, how they heat and light their homes, their everyday items like toilets and stoves, and in total more than 130 different aspects of their daily lives. We could fill a whole book with images showing the striking similarities between the lives of people living on the same incomes in different countries, and the huge differences in how people live within countries. We have over 40,000 photos.*

What the photos make clear is that the main factor that affects how people live is not their religion, their culture, or the country they live in, but their income.

TOOTHBRUSHES

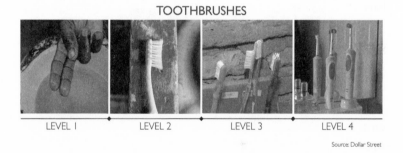

| LEVEL 1 | LEVEL 2 | LEVEL 3 | LEVEL 4 |

Source: Dollar Street

Here are some toothbrushes from families with different income levels. On Level 1 you brush with your finger or a stick. On Level 2 you get a plastic toothbrush. On Level 3 you get one each. And Level 4 you are already familiar with.

The bedrooms (or kitchens or living rooms) of families living on Level 4 look very similar in the United States, Vietnam, Mexico, South Africa, or anywhere else in the world.

---

* Visit Dollar Street here: www.dollarstreet.org.

BEDS ON LEVEL 4

Typical beds from homes with more than $32/day across the world.

Source: Dollar Street

The way a family living on Level 2 in China stores and prepares food looks very similar to the way a family living on Level 2 in Nigeria stores and prepares food.

LEVEL 2 STOVES: OPEN FIRES

Source: Dollar Street

In fact, when you are one of the 3 billion people living on Level 2, whether you live in the Philippines, Colombia, or Liberia, the basic facts about your life are quite similar.

Your house has a patchwork roof, so if it's raining you might well get wet and cold.

LEVEL 2 ROOFS: PATCHWORKS

Philippines   Colombia   Liberia

Source: Dollar Street

When you go to the toilet in the morning it is smelly and full of flies, but at least there are some walls to give you some privacy.

LEVEL 2 TOILETS: PITS

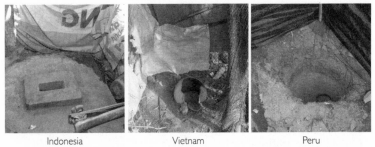

Indonesia   Vietnam   Peru

Source: Dollar Street

You eat the same for almost every meal, every day of every week. You dream about food that is more varied and more delicious.

The light flickers because the electricity is unstable. You have to rely on moonlight on the nights when the power is out. You secure the door using a padlock.

When you go to bed in the evening you might brush your teeth with the same toothbrush as the rest of the family. You dream about the day when you don't have to share your toothbrush with Grandma anymore.

In the media, we see photos of everyday life on Level 4 and crisis on the other levels all the time. Google *toilet, bed,* or *stove.* You will get images from Level 4. If you want to see what everyday life is like on the other levels, Google won't help.

Question Your Categories

It will be helpful to you if you always assume your categories are misleading. Here are five powerful ways to keep questioning your favorite categories: look for differences within and similarities across groups; beware of "the majority"; beware of exceptional examples; assume you are not "normal"; and beware of generalizing from one group to another.

Look for Differences Within Groups and Similarities Across Groups

Country stereotypes simply fall apart when you look at the huge differences within countries and the equally huge similarities between countries on the same income level, independent of culture or religion.

Remember the similarities between the cooking pots of families on Level 2 in Nigeria and China? If you saw just the picture from China you would probably think, "Oh, that's how they heat water in China. In an iron pot on a tripod over a fire. That's their culture." No. It is a common way to heat water on Level 2, all over the world. It's a question of income. And in China, as elsewhere, people also cook in several other ways, depending not on their "culture" but on their income level.

When someone says that an individual did something because they

DIFFERENCES IN HEALTH AND WEALTH IN AFRICA

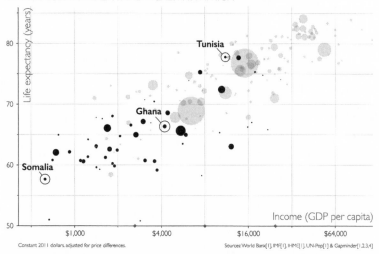

Constant 2011 dollars, adjusted for price differences.　　　　　Sources: World Bank[1], IMF[1], IHME[1], UN-Pop[1] & Gapminder[1,2,3,4]

belong to some group—a nation, a culture, a religion—take care. Are there examples of different behavior in the same group? Or of the same behavior in other groups?

Africa is a huge continent of 54 countries and 1 billion people. In Africa we find people living at every level of development: in the bubble chart above I have highlighted all the African countries. Look at Somalia, Ghana, and Tunisia. It makes no sense to talk about "African countries" and "Africa's problems" and yet people do, all the time. It leads to ridiculous outcomes like Ebola in Liberia and Sierra Leone affecting tourism in Kenya, a 100-hour drive across the continent. That is farther than London to Tehran.

Beware of "The Majority"

When someone says that a majority of a group has some property it can sound like most of them have something in common. Remember that *majority* just means more than half. It could mean 51 percent. It could mean 99 percent. If possible, ask for the percentage.

For example, here's a fact: In all countries in the world, a majority of women say their needs for contraceptives are met. What does that tell us? Does it mean nearly everyone? Or does it mean a little over half? The reality differs widely from one country to another. In China and France, an impressive 96 percent of women say their needs for contraception are being met. Just below that, at 94 percent, are the United Kingdom, South Korea, Thailand, Costa Rica, Nicaragua, Norway, Iran, and Turkey. But in Haiti and Liberia, "the majority" means just 69 percent, and in Angola it means only 63 percent.

Beware of Exceptional Examples
Beware of exceptional examples used to make a point about a whole group. Chemophobia, the fear of chemicals, is driven by generalizations from a few vivid but exceptional examples of harmful substances. Some people become frightened of all "chemicals." But remember that everything is made from chemicals, all "natural" things and all industrial products. Here are some of my favorites that I would rather not live without: soap, cement, plastic, washing detergent, toilet paper, and antibiotics. If someone offers you a single example and wants to draw conclusions about a group, ask for more examples. Or flip it over: i.e., ask whether an opposite example would make you draw the opposite conclusion. If you are happy to conclude that all chemicals are unsafe on the basis of one unsafe chemical, would you be prepared to conclude that all chemicals are safe on the basis of one safe chemical?

Assume You Are Not "Normal" and Other People Are Not Idiots
To avoid getting your leg crushed in an elevator and other bad mistakes, stay open to the possibility that your experience might not be "normal." Be cautious about generalizing from Level 4 experiences

to the rest of the world. Especially if it leads you to the conclusion that other people are idiots.

If you were to visit Tunisia, a country where you find people living on every level from 1 to 4, you might come across houses that were half-built—like this one, belonging to the Salhi family, who live in the capital, Tunis. You might conclude that Tunisians were lazy or disorganized.

Source: Dollar Street

You can visit the Salhi family on Dollar Street and see how they live. Mabrouk is 52 years old and is a gardener. His wife, Jamila, is 44 years old and runs a home-based bakery. Most of their neighbors have similar half-built second floors on their houses. You see this everywhere on Levels 2 and 3 across the world. In Sweden, if someone built their house like that, we would think they had a severe planning problem, or maybe the builders had run away. But you can't generalize from Sweden to Tunisia.

The Salhis, and many others living in similar circumstances, have found a brilliant way to solve several problems at once. On Levels 2 and 3, families often do not have access to a bank to put their savings and cannot get a loan. So, to save up to improve their home, they must pile up money. Money, though, can be stolen or lose its value through inflation. So, instead, whenever they can afford them, the Salhis buy actual bricks, which won't lose their value. But there is no space inside to store the bricks and the bricks might get stolen if they are left in a pile outside. Better to add the bricks to the house as you buy them. Thieves can't steal them. Inflation won't change their value. No one needs to check your credit rating. And over 10 or 15 years you are slowly building your family a better home. Instead of assuming that the Salhis are lazy or disorganized, assume they are smart and ask yourself, How can this be such a smart solution?

Beware of Generalizing from One Group to Another

I once used to believe and promote a fatally incorrect generalization that cost 60,000 lives. Some of those lives could have been saved if the public health community had been keener to question its misleading generalizations.

One evening in 1974, I was shopping for bread at a supermarket in a small Swedish town when I suddenly discovered a baby in a life-threatening situation. In a stroller in the bread aisle. The mother had turned her back and was busy deciding which loaf to buy. An untrained eye couldn't see the danger, but fresh out of medical school, I heard my alarm bells go off. I restrained myself from running, to not scare the mother. Instead I walked over to the stroller as quickly as I could and I lifted up the baby, who was asleep on his back. I turned him over and put him down on his tummy. The little fellow didn't even wake up.

The mother turned toward me with a loaf in her hand, ready to attack. I quickly explained to her that I was a physician and I told her about the so-called sudden infant death syndrome and the new public health advice to parents: not to put sleeping babies on their backs due to the risk of suffocation from vomiting. Now her baby was safe. The mother was both scared and comforted. On trembling legs she continued her shopping. Proudly I completed my own purchases, unaware of my huge mistake.

During the Second World War and the Korean War, doctors and nurses discovered that unconscious soldiers stretchered off the battlefields survived more often if they were laid on their fronts rather than on their backs. On their backs, they often suffocated on their own vomit. On their fronts, the vomit could exit and their airways remained open. This observation saved many millions of lives, not just of soldiers. The "recovery position" has since become a global best practice, taught in every first-aid course on the planet. (The rescue workers saving lives after the 2015 earthquake in Nepal had all learned it.)

But a new discovery can easily be generalized too far. In the 1960s, the success of the recovery position inspired new public health advice, against most traditional practices, to put babies to sleep on their tummies. As if any helpless person on their back needed *just* the same help.

The mental clumsiness of a generalization like this is often difficult to spot. The chain of logic seems correct. When seemingly impregnable logic is combined with good intentions, it becomes nearly impossible to spot the generalization error. Even though the data showed that sudden infant deaths went up, not down, it wasn't until 1985 that a group of pediatricians in Hong Kong actually suggested that the prone position might be the cause. Even then, doctors in Europe didn't pay much attention. It took Swedish authorities another seven years to accept their mistake and reverse the policy. Unconscious soldiers were dying on their backs when they vomited. Sleeping babies, unlike unconscious soldiers, have fully functioning reflexes and turn to the side if they vomit while on their backs. But on their tummies, maybe some babies are not yet strong enough to tilt their heavy heads to keep their airways open. (The reason the prone position is more dangerous is still not fully understood.)

It's difficult to see how the mother in the bread aisle could have realized I was putting her baby at risk. She could have asked me for evidence. I would have told her about the unconscious soldiers. She could have asked, "But dear doctor, is that really a valid generalization? Isn't a sleeping baby very different from an unconscious soldier?" Even if she had put this to me, I strongly doubt I would have been able to think it through.

With my own hands, over a decade or so, I turned many babies from back to tummy to prevent suffocation and save lives. So did many other doctors and parents throughout Europe and the United States, until the advice was finally reversed, 18 months after the Hong Kong study was published. Thousands of babies died because of a sweeping generalization, including some during the months when the evidence was already available. Sweeping generalizations can easily hide behind good intentions.

I can only hope that the baby in the bread aisle survived. And I can only hope that people are willing to learn from this huge public health mistake in modern times. We must all try hard not to generalize across incomparable groups. We must all try hard to discover the hidden sweeping generalizations in our logic. They are very difficult to discover. But when presented with new evidence, we must always be ready to question our previous assumptions and reevaluate and admit if we were wrong.

Factfulness

Factfulness is . . . recognizing when a category is being used in an explanation, and remembering that categories can be misleading. We can't stop generalization and we shouldn't even try. What we should try to do is to avoid generalizing incorrectly.

To control the generalization instinct, **question your categories.**

- **Look for differences *within* groups.** Especially when the groups are large, look for ways to split them into smaller, more precise categories. **And . . .**
- **Look for similarities *across* groups.** If you find striking similarities between different groups, consider whether your categories are relevant. **But also . . .**
- **Look for *differences* across groups.** Do not assume that what applies for one group (e.g., you and other people living on Level 4 or unconscious soldiers) applies for another (e.g., people not living on Level 4 or sleeping babies).
- **Beware of "the majority."** The majority just means more than half. Ask whether it means 51 percent, 99 percent, or something in between.
- **Beware of vivid examples.** Vivid images are easier to recall but they might be the exception rather than the rule.
- **Assume people are not idiots.** When something looks strange, be curious and humble, and think, In what way is this a smart solution?

CHAPTER SEVEN

THE DESTINY INSTINCT

About rocks that move and what
Grandpa never talked about

Snowballs in Hell

Not long ago I was invited to the five-star Balmoral Hotel in Edinburgh to present to a gathering of capital managers and their wealthiest clients. As I set up my equipment in the magnificent high-ceilinged ballroom, I couldn't help feeling a bit small, and I asked myself why a wealthy financial institution would want its clients to hear from a Swedish professor of public health. I had been carefully briefed weeks earlier, but to feel sure, I asked the conference organizer again as I got onstage for a final rehearsal. He had a straightforward explanation. He was having a hard time making his clients understand that the most profitable investments were no longer to be found in European capitals boasting medieval castles and cobbled streets, but in the emerging markets of Asia and Africa. "Most of our clients," he said, "are unable to see or accept the ongoing progress in many African countries. In their minds, Africa is a continent that will never improve. I want your moving charts to change their static view of the world."

My lecture seemed to go well. I showed how Asian countries like South Korea, China, Vietnam, Malaysia, Indonesia, the Philippines, and Singapore, which had surprised the world with their economic progress over the past decades, actually had made steady social progress during the decades before their economic growth. I showed how the same process was now unfolding in parts of Africa. I said that the best places to invest right now were probably those African countries that had just seen decades of rapid improvements in education and child survival. I mentioned Nigeria, Ethiopia, and Ghana. The audience listened hard, eyes wide, and asked some good questions.

Afterward, as I was packing away my laptop, a gray-haired man in a lightly checked three-piece suit walked slowly up to the stage, smiled sweetly, and said, "Well, I saw your numbers and I heard what you said, but I'm afraid there's not a snowball's chance in hell that Africa will make it. I know because I served in Nigeria. It's their culture, you know. It will not allow them to create a modern society. Ever. EV-ER." I opened my mouth, but before I had figured out a fact-based reply, he had already given my shoulder a little pat and wandered off to find a cup of tea.

The Destiny Instinct

The destiny instinct is the idea that innate characteristics determine the destinies of people, countries, religions, or cultures. It's the idea that things are as they are for ineluctable, inescapable reasons: they have always been this way and will never change. This instinct makes us believe that our false generalizations from chapter 6, or the tempting gaps from chapter 1, are not only true, but fated: unchanging and unchangeable.

It is easy to see how this instinct would have served an evolutionary purpose. Historically, humans lived in surroundings that didn't change much. Learning how things worked and then assuming they would continue to work that way rather than constantly reevaluating was probably an excellent survival strategy.

It's also easy to understand how claiming a particular destiny for your group can come in useful in uniting that group around a supposedly never-changing purpose, and perhaps creating a sense of superiority over other groups. Such ideas must have been important for powering tribes, chiefdoms, nations, and empires. But today, this instinct to see things as unchanging, this instinct not to update our knowledge, blinds us to the revolutionary transformations in societies happening all around us.

Societies and cultures are not like rocks, unchanging and unchangeable. They move. Western societies and cultures move, and non-Western societies and cultures move—often much faster. It's just that all but the fastest cultural changes—the spread of the internet, smartphones, and social media, for example—tend to happen just a bit too slowly to be noticeable or newsworthy.

A common expression of the destiny instinct is my Edinburgh gentleman's idea that Africa will always be a basket case and will never catch up with Europe. Another is that the "Islamic world" is fundamentally different from the "Christian world." This or that religion or continent or culture or nation will (or must) never change, because of its traditional and unchanging "values": again and again, it's the same idea in different costumes. At first sight there appears to be some analysis going on. On closer inspection, our instincts have often fooled us. These lofty statements are often simply feelings disguised as facts.

FACT QUESTION 10

Worldwide, 30-year-old men have spent 10 years in school, on average. How many years have women of the same age spent in school?

☐   A: 9 years
☐   B: 6 years
☐   C: 3 years

By now I hope you have worked out that the safest thing to do in this book is to pick the most positive answer. Thirty-year-old women have on average spent nine years in school, just one year less than the men.

Many of my fellow Europeans have a snobbish self-regard built on an illusion of a European culture that is superior, not only to African and Asian cultures, but also to US consumer culture. When it comes to drama, though, I wonder who consumes the most. Twenty-six percent of the US public picked the right answer, compared with 13 percent in Spain and Belgium, 10 percent in Finland, and just 8 percent in Norway.

FACT QUESTION 10 RESULTS: percentage who answered correctly.
Worldwide, 30-year-old men have spent 10 years in school, on average.
How many years have women of the same age spent in school? (Correct answer: 9 years.)

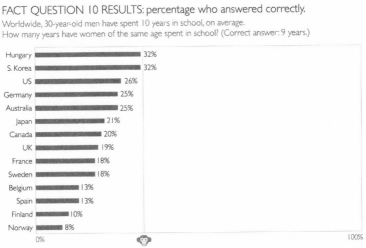

Sources: Ipsos MORI[1] & Novus[1]

The question is about gender inequality, which is currently discussed in the Scandinavian media on a daily basis. We see constant examples of the brutal violence committed against women out there, mostly elsewhere, in the rest of the world, as well as reports from places like Afghanistan, where many, many girls are out of school. These images confirm a popular idea in Scandinavia that gender equality elsewhere has not improved—that most other cultures are stuck.

How the Rocks Move

Cultures, nations, religions, and people are not rocks. They are in constant transformation.

Africa Can Catch Up

The idea that Africa is destined to remain poor is very common but often seems to be based on no more than a feeling. If you like your opinions to be based on facts, this is what you need to know.

Yes, Africa is lagging behind other continents, on average. The average lifespan of a newborn baby in Africa today is 65 years. That's a staggering 17 years less than a baby born today in Western Europe.

But, first of all, you know how misleading averages can be, and the differences within Africa are immense. Not all African countries are lagging the world. Five large African countries—Tunisia, Algeria, Morocco, Libya, and Egypt—have life expectancies above the world average of 72 years. They are where Sweden was in 1970.

Those despairing for Africa may not be convinced by this example. They may say that these are all Arab countries on the north coast of Africa and therefore not the Africa they had in mind. When I was young, these countries were certainly seen as sharing Africa's destiny. It is only since they made progress that they have been held to be exceptional. For the sake of argument, though, let's put these North African countries to one side and look at Africa south of the Sahara.

In the last 60 years the African countries south of the Sahara almost all went from being colonies to being independent states. Over that time, these countries expanded their education, electricity, water, and sanitation infrastructures at the same steady speed as that achieved by the countries of Europe when they went through their own miracles. And each of the 50 countries south of the Sahara reduced its child mortality faster than Sweden ever did. How can that not be counted as incredible progress?

Perhaps because though things are much better, they are still bad. If you look for poor people in Africa, of course you will find them.

But there was extreme poverty in Sweden 90 years ago too. And when I was young, just 50 years ago, China, India, and South Korea were all way behind where sub-Saharan Africa is today in most ways, and Asia's destiny was supposed then to be exactly what Africa's destiny is supposed to be now: "They will never be able to feed 4 billion people."

Roughly half a billion people in Africa today are stuck in extreme poverty. If it is their destiny to stay that way, then there must be something unique about this particular group of poor people compared with the billions across the world, including in Africa, who have already escaped extreme poverty. I don't think there is.

I think the last to leave extreme poverty will be the poorest farmers stuck on really low-yield soils and surrounded by or close to conflicts. That probably accounts today for 200 million people, just over half of whom live in Africa. For sure they have an extraordinarily difficult time ahead of them—not because of their unchanging and unchangeable culture, but because of the soil and the conflicts.

But I hold out hope even for these poorest and most unfortunate people in the world, because this is exactly how hopeless extreme poverty has always seemed. During their terrible famines and conflicts, progress in China, Bangladesh, and Vietnam seemed impossible. Today these countries probably produced most of the clothes in your wardrobe. Thirty-five years ago, India was where Mozambique is today. It is fully possible that within 30 years Mozambique will transform itself, as India has done, into a country on Level 2 and a reliable trade partner. Mozambique has a long, beautiful coast on the Indian Ocean, the future center of global trade. Why should it not prosper?

Nobody can predict the future with 100 percent certainty. I'm not convinced it will happen. But I am a possibilist and these facts convince me: it is possible.

The destiny instinct makes it difficult for us to accept that Africa can catch up with the West. Africa's progress, if it is noticed at all, is seen as an improbable stroke of good fortune, a temporary break from its impoverished and war-torn destiny.

The same destiny instinct also seems to make us take continuing Western progress for granted, with the West's current economic stagnation portrayed as a temporary accident from which it will soon recover. For years after the global crash of 2008, the International Monetary Fund continued to forecast 3 percent annual economic growth for countries on Level 4. Each year, for five years, countries on Level 4 failed to meet this forecast. Each year, for five years, the IMF said, "Next year it will get back on track." Finally, the IMF realized that there was no "normal" to go back to, and it downgraded its future growth expectations to 2 percent. At the same time the IMF acknowledged that the fast growth (above 5 percent) during those years had instead happened in countries on Level 2, like Ghana, Nigeria, Ethiopia, and Kenya in Africa, and Bangladesh in Asia.

Why does this matter? One reason is this: the IMF forecasters' worldview had a strong influence on where your retirement funds were invested. Countries in Europe and North America were expected to experience fast and reliable growth, which made them attractive to investors. When these forecasts turned out to be wrong, and when these countries did not in fact grow fast, the retirement funds did not grow either. Supposedly low-risk/high-return countries turned out to be high-risk/low-return countries. And at the same time African countries with great growth potential were being starved of investment.

Another reason it matters, if you work for a company based in the old "West," is that you are probably missing opportunities in the largest expansion of the middle-income consumer market in history, which is taking place right now in Africa and Asia. Other, local brands are already establishing a foothold, gaining brand recognition, and spreading throughout these continents, while you are still waking up to what is going on. The Western consumer market was just a teaser for what is coming next.

Babies and Religions

At the end of my opening lecture in my 1998 course on global health, most students headed for the coffee machine but one remained behind. I saw her wander slowly toward the front of the room with tears in her eyes, then, when she understood that I had noticed her, she stopped, flipped her face away, and looked out the window. She was obviously moved. I expected her to share with me a sad personal problem that was going to impede her participation in the course. Before I could say anything comforting she turned around, gained control over her emotions, and in a steady voice said something completely unexpected:

"My family is from Iran. What you just said about the fast improvements in health and education in Iran was the first positive thing I've heard anyone from Sweden ever say about the Iranian people."

My student said this to me in perfect Swedish with a clear Stockholm accent: she had obviously lived in Sweden her whole life. I was stunned. All I had done was to briefly show UN data for Iran on the increase in life expectancy and decrease in babies per woman. I had mentioned too that it was quite an achievement—actually the fastest drop ever, from more than six babies per woman in 1984 down to fewer than three babies per woman just 15 years later.

AVERAGE BABIES PER WOMAN FROM 1800 TO TODAY

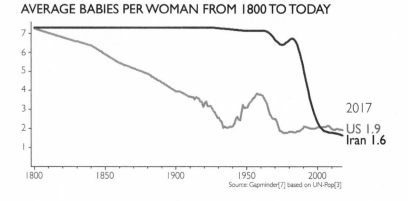

Source: Gapminder[7] based on UN-Pop[3]

It was one of several little-known examples I had shown of fast changes in middle-income countries in the 1990s.

"That can't be true," I said.

"It is. You said that the fast fall in the number of babies per woman in Iran is a reflection of improvements in health and education, especially for Iranian women. You also rightly said that most young Iranians now have modern values about family size and use contraception. I have never heard anyone in Sweden say anything even close to that. Even highly educated Swedes seem completely unaware of the changes that have taken place. The improvements. The modernity. They think Iran is on the same level as Afghanistan."

The fastest drop in babies per woman in world history went completely unreported in the free Western media. Iran—home in the 1990s to the biggest condom factory in the world, and boasting a compulsory pre-marriage sex education course for both brides and grooms—has a highly educated population with excellent access to an advanced public health-care system. Couples use contraception to achieve small families and have access to infertility clinics if they struggle to conceive. At least that was the case when I visited such a clinic in Tehran in 1990, hosted by the enthusiastic Professor Malek-Afzali, who designed Iran's family planning miracle.

How many people in the West would guess that women in Iran today decide to have fewer babies than women in either the United States or Sweden? Do we Westerners love free speech so much that it makes us blind to any progress in a country whose regime does not share our love? It is, at least, clear that a free media is no guarantee that the world's fastest cultural changes will be reported.

Almost every religious tradition has rules about sex, so it is easy to understand why so many people assume that women in some religions give birth to more children. But the link between religion and the number of babies per woman is often overstated. There is, though, a strong link between income and number of babies per woman.

Back in 1960 this didn't seem so obvious. In 1960, there were 40 countries where women had fewer than 3.5 babies on average, and they were all Christian-majority countries, except Japan. It appeared that to have few babies, you had either to be Christian or Japanese. (A bit more reflection even at this stage would have suggested some problems with this line of thought: in many Christian-majority countries, like Mexico and Ethiopia, women also had big families.)

How does it look today? In the bubble graphs on the next page, I have divided the world into three groups based on religion: Christian, Muslim, or other. I have then shown babies per woman and income for each group. As usual the size of the bubble reflects the size of the population. Look how Christian populations are spread out on all income levels. Look how the Christian populations on Level 1 have many more babies. Now look at the other two graphs. The pattern is very similar: regardless of religion, women have more children if they live in extreme poverty on Level 1.

WITH HIGHER INCOME COMES FEWER BABIES

All countries split into religious groups, 2017. Bubble size shows number of people.

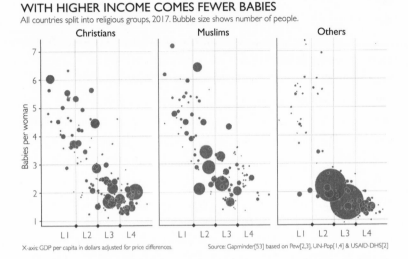

X-axis: GDP per capita in dollars adjusted for price differences.   Source: Gapminder[53] based on Pew[2,3], UN-Pop[1,4] & USAID-DHS[2]

Today, Muslim women have on average 3.1 children. Christian women have 2.7. There is no major difference between the birth rates of the great world religions.

In almost every bedroom, across continents, cultures, and religions—in the United States, Iran, Mexico, Malaysia, Brazil, Italy, China, Indonesia, India, Colombia, Bangladesh, South Africa, Libya, you name it—couples are whispering into each other's ears their dreams for their future happy families.

Everyone's Talking About Sex

Exaggerated claims that people from this religion or that religion have bigger families are one example of how people tend to claim that certain values or behaviors are culture-specific, unchanging and unchangeable.

It's just not true. Values change all the time.

Take my lovely home country, Sweden. We Swedes are known for being quite liberal and open about sex and contraception, aren't we? Yet this hasn't always been our culture. These haven't always been our values.

In my own living memory, Swedish values around sex were extremely conservative. My father's father, Gustav, for example, was born as Sweden was leaving Level 2 and was, I believe, a quite typical Swedish man of his generation. He was extremely proud of his large family of seven children; he never changed a diaper, cooked food, or cleaned the house; and he absolutely would not talk about sex or contraception. His oldest daughter supported the brave feminists who illegally started advocating the use of condoms in the 1930s. But when she approached her father after the birth of his seventh child, wanting to discuss contraception, this kind, calm man got very angry and refused to talk. His values were traditional and patriarchal. But they were not adopted by the next generation. Swedish culture changed. (By the way, he also disliked books and refused to use a telephone.)

A woman's right to an abortion is supported by just about everyone in Sweden today. Strong support for women's rights in general has become part of our culture. My students' jaws drop when I tell them how different things were when I was a student in the 1960s. Abortion in Sweden was still, except on very limited grounds, illegal. At the university, we ran a secret fund to pay for women to travel abroad to get safe abortions. Jaws drop even further when I tell the students where these young pregnant students traveled to: Poland. Catholic Poland. Five years later, Poland banned abortion and Sweden legalized it. The flow of young women started to go the other way. The point is, it was not always so. The cultures changed.

I come across the values of stubborn old men like my grandfather Gustav all the time when I travel in Asia. For example, in South Korea and Japan, many wives are still expected to take care of their husband's parents, as well as taking full responsibility for the care of any children. I have encountered many men who are proud of these "Asian values," as they call them. I have had conversations with many women too, who see it differently. They find this culture unbearable and tell me these values make them less interested in getting married.

The Idea of a Husband

At a banking conference in Hong Kong, I was seated at dinner next to a brilliant young banker. She was 37 years old and enjoying a very successful career, and she taught me many things over dinner about current issues and trends in Asia. Then we started talking about our personal lives. "Do you plan to have a family?" I asked. I didn't mean to be rude: we Swedes (nowadays) like to talk about these things. And she had no problem with my honest question. She smiled and looked over my shoulder at the sun setting over the bay. She said, "I am thinking about children every day." Then she looked me straight in the eye. "It's the idea of a husband I can't stand."

I try to comfort these women, to convince them that things will change. I recently gave a lecture to 400 young women at the Asian University for Women in Bangladesh. I told them about how and why cultures are always transforming, how escape from extreme poverty and women's access to education and contraception have led to more pillow talk and fewer children. It was a very emotional lecture. The young women in colorful hijabs smiled with their whole faces.

Afterward, the Afghan students wanted to tell me about their country. They told me these changes were already slowly happening even in Afghanistan. "Despite the war, despite the poverty," they told me, "many of us young people are planning a modern life. We are Afghans, we are Muslim women. And we want a man just like you describe, a man who listens and plans together with us, and then we want two children who go to school."

The macho values that are found today in many Asian and African countries, these are not Asian values, or African values. They are not Muslim values. They are not Eastern values. They are patriarchal values like those found in Sweden only 60 years ago, and with social and economic progress they will vanish, just as they did in Sweden. They are not unchangeable.

How to Control the Destiny Instinct

How can we help our brains to see that rocks move; that the way things are now is neither how they have always been nor how they are always meant to be?

Slow Change Is Not No Change

Societies and cultures are in constant movement. Even changes that seem small and slow add up over time: 1 percent growth each year seems slow but it adds up to a doubling in 70 years; 2 percent growth each year means doubling in 35 years; 3 percent growth each year means doubling in 24 years.

In the third century BC, the world's first nature reserve was created by King Devanampiya Tissa in Sri Lanka when he declared a piece of forest to be officially protected. It took more than 2,000 years for a European, in West Yorkshire, to get a similar idea, and another 50 years before Yellowstone National Park was established in the United States. By the year 1900, 0.03 percent of the Earth's land surface was protected. By 1930 it was 0.2 percent. Slowly, slowly, decade by decade, one forest at a time, the number climbed. The annual increase was absolutely tiny, almost imperceptible. Today a stunning 15 percent of the Earth's surface is protected, and the number is still climbing.

To control the destiny instinct, don't confuse slow change with no change. Don't dismiss an annual change—even an annual change of only 1 percent—because it seems too small and slow.

Be Prepared to Update Your Knowledge

It's relaxing to think that knowledge has no sell-by date: that once you have learned something, it stays fresh forever and you never have to learn it again. In the sciences like math and physics, and in the arts, that is often true. In those subjects, what we all learned at school $(2+2=4)$ is probably still good. But in the social sciences, even the most basic knowledge goes off very quickly. As with milk or vegetables, you have to keep getting it fresh. Because everything changes.

I have been caught out by this even in my own work. Thirteen years after I first asked them, we planned to rerun my very first chimpanzee questions from 1998 to see whether people's knowledge had improved. In these questions, I showed five pairs of countries and asked which country in each pair had the highest child mortality rate. Back in 1998, my Swedish students had answered incorrectly because they couldn't imagine that Asian countries were better than European countries.

When we pulled the questions up, after only 13 years, we realized that it was going to be impossible to rerun the test because the correct answers had changed. Because the world had changed. How illustrative was this? Even Gapminder's own fact questions had become outdated.

To control the destiny instinct, stay open to new data and be prepared to keep freshening up your knowledge.

Talk to Grandpa

If you are tempted to claim that values are unchanging, try comparing your own with those of your parents, or your grandparents—or your children or your grandchildren. Try getting hold of public opinion polls for your country from 30 years ago. You will almost certainly see radical change.

Collect Examples of Cultural Change

People often tilt their heads and say "it's our culture" or "it's their culture," which gives the impression that it has always been that way and always will be. Then turn your head around and look for some counterexamples. We already discovered that Swedes didn't always talk about sex. Here are a couple of others.

Many Swedes think of the United States as having very conservative values. But look at how quickly attitudes to homosexuality have changed. In 1996, a minority of 27 percent supported same-sex marriage. Today that number is 72 percent and rising.

Some Americans think of Sweden as a socialist country, but values can change. A few decades ago Sweden carried out what might be the most drastic deregulation ever of a public school system and now allows fully commercial schools to compete and make profits (a brave capitalist experiment).

I Don't Have Any Vision

I started this chapter with a story about a well-dressed ignorant man who didn't have sufficient vision to see what was possible in Africa. I want to end with something similar. (Spoiler alert: the ignorant man *this* time is me.)

On May 12, 2013, I had the great privilege of addressing 500 women leaders from across the continent at an African Union conference called "The African Renaissance and Agenda for 2063." What an enormous honor, what a thrill. It was the lecture of my life. In my 30-minute slot in the Plenary Hall of the African Union's beautiful headquarters in Addis Ababa, I summarized decades of research on female small-scale farmers and explained to these powerful decision makers how extreme poverty could be ended in Africa within 20 years.

Nkosazana Dlamini-Zuma, the chairperson of the African Union Commission, sat right in front of me and seemed to be listening attentively. Afterward, she came up and thanked me and I asked her what she thought of my performance. Her answer was a shock.

"Well," she said, "the graphics were nice, and you are good at talking, but you don't have any vision." Her tone was kind, which made what she was saying even more shocking to me.

"What?! You think I lack vision?" I asked in offended disbelief. "But I said that extreme poverty in Africa could be history within 20 years."

Nkosazana's response came in a low voice and she spoke without emotions or gestures. "Oh, yes, you talked about eradicating extreme poverty, which is a beginning, but you stopped there. Do you think Africans will settle with getting rid of extreme poverty and be happy living in only ordinary poverty?" She put a firm hand on my arm and looked at me without anger but also without a smile. I saw a strong will to make me understand my shortcomings. "As a finishing remark you said that you hoped your grandchildren would come as tourists to Africa and travel on the new high-speed trains we plan to build. What kind of a vision is that? It is the same old European vision." Nkosazana looked me straight in my eyes. "It is *my* grandchildren who are going to visit *your* continent and travel on *your* high-speed trains and visit that exotic ice hotel I've heard you have up in northern Sweden. It is going to take a long time, we know that. It is going to take lots of wise decisions and large investments. But my 50-year vision is that Africans will be welcome tourists in Europe and not unwanted refugees." Then she broke into a broad, warm smile. "But the graphics were really nice. Now let's go and have some coffee."

Over coffee I reflected on my mistake. I remembered a conversation from 33 years earlier with my first African friend, the Mozambican mining engineer Niherewa Maselina. He had looked at me with that same face. I was working as a doctor in Nacala in Mozambique, and Niherewa had come with us on a family outing to the beach. The coast in Mozambique is unbelievably beautiful and was still hardly exploited and we used to be almost alone there at the weekends. When I saw that there were 15 or 20 families on the mile-long stretch of sand I said, "Oh, what a shame there are so many families on the beach today." Niherewa grabbed my arm, just as Nkosazana was to do years later, and said, "Hans. My reaction is the opposite. I feel great pain and sadness seeing this beach. Look at the city there in the distance. Eighty thousand people live there, which means 40,000 children. It's the weekend. And only 40 of them made it to the beach. One in one thousand. When I got my mining education in East Germany, I went to the beaches of Rostock at the weekend, and they were full. Thousands of children having a wonderful time. I want Nacala to be like Rostock. I want all children to go to the beach on a Sunday instead of working in their parents' fields or sitting in the slums. It will take a long time, but that is what I want." Then he let go of my arm and helped my children to get their swimming gear out of the car.

Thirty-three years later, addressing the African Union after a professional lifetime of collaboration with African scholars and institutions, I was absolutely convinced that I shared their great vision. I thought I was one of the few Europeans who could see what change was possible. But after delivering the most cherished lecture of my life, I realized that I was still stuck in an old, static, colonial mind-set. In spite of all that my African friends and colleagues had taught me over the years, I was still not really imagining "they" could ever catch up with "us." I was still failing to see that all people, families, children will struggle hard to achieve just that, so they can also go to the beach.

Factfulness

Factfulness is . . . recognizing that many things (including people, countries, religions, and cultures) appear to be constant just because the change is happening slowly, and remembering that even small, slow changes gradually add up to big changes.

To control the destiny instinct, **remember slow change is still change.**

- **Keep track of gradual improvements.** A small change every year can translate to a huge change over decades.
- **Update your knowledge.** Some knowledge goes out of date quickly. Technology, countries, societies, cultures, and religions are constantly changing.
- **Talk to Grandpa.** If you want to be reminded of how values have changed, think about your grandparents' values and how they differ from yours.
- **Collect examples of cultural change.** Challenge the idea that today's culture must also have been yesterday's, and will also be tomorrow's.

CHAPTER EIGHT

THE SINGLE PERSPECTIVE INSTINCT

Why governments should not be mistaken for nails
and why shoes and bricks sometimes tell you
more than numbers

Who Can We Trust?

Forming your worldview by relying on the media would be like forming your view about me by looking only at a picture of my foot. Sure, my foot is part of me, but it's a pretty ugly part. I have better parts. My arms are unremarkable but quite fine. My face is OK. It isn't that the picture of my foot is deliberately lying about me. But it isn't showing you the whole of me.

Where, then, shall we get our information from if not from the media? Who can we trust? How about experts? People who devote their working lives to understanding their chosen slice of the world? Well, you have to be very careful here too.

The Single Perspective Instinct

We find simple ideas very attractive. We enjoy that moment of insight, we enjoy feeling we really understand or know something. And it is easy to take off down a slippery slope, from one attention-grabbing simple idea to a feeling that this idea beautifully explains, or is the beautiful solution for, lots of other things. The world becomes simple. All problems have a single cause—something we must always be completely against. Or all problems have a single solution—something we must always be for. Everything is simple. There's just one small issue. We completely misunderstand the world. I call this preference for single causes and single solutions the single perspective instinct.

For example, the simple and beautiful idea of the free market can lead to the simplistic idea that all problems have a single cause—government interference—which we must always oppose; and that the solution to all problems is to liberate market forces by reducing taxes and removing regulations, which we must always support.

Alternatively, the simple and beautiful idea of equality can lead to the simplistic idea that all problems are caused by inequality, which we should always oppose; and that the solution to all problems is redistribution of resources, which we should always support.

It saves a lot of time to think like this. You can have opinions and answers without having to learn about a problem from scratch and you can get on with using your brain for other tasks. But it's not so useful if you like to understand the world. Being always in favor of or always against any particular idea makes you blind to information that doesn't fit your perspective. This is usually a bad approach if you like to understand reality.

Instead, constantly test your favorite ideas for weaknesses. Be humble about the extent of your expertise. Be curious about new information that doesn't fit, and information from other fields. And rather than talking only to people who agree with you, or collecting examples that fit your ideas, see people who contradict you, disagree with you, and put forward different ideas as a great resource for understanding the world. I have been wrong about the world so many times. Sometimes, coming up against reality is what helps me see my mistakes, but often it is talking to, and trying to understand, someone with different ideas.

If this means you don't have time to form so many opinions, so what? Wouldn't you rather have few opinions that are right than many that are wrong?

I have found two main reasons why people often focus on a single perspective when it comes to understanding the world. The obvious one is political ideology, and I will come to that later in this chapter. The other is professional.

The Professionals: Experts and Activists

I love subject experts, and as we all must do, I rely heavily on them to understand the world. When I know, for example, that all population experts agree that population will stop growing somewhere between 10 billion and 12 billion, then I trust that data. When I know, for example, that historians, paleodemographers, and archeologists have all concluded that until 1800, women had on average five or more children but only two survived, I trust that data. When I know that economists disagree about what causes economic growth, that is extremely useful too, because it tells me I must be careful: probably there is not enough useful data yet, or perhaps there is no simple explanation.

I love experts, but they have their limitations. First, and most obviously, experts are experts only within their own field. That can be difficult for experts (and we are all experts in something) to admit. We like to feel knowledgeable and we like to feel useful. We like to feel that our special skills make us generally better.

But . . .

Highly numerate people (like the super-brainy audience at the Amazing Meeting, an annual gathering of people who love scientific reasoning) score just as badly on our fact questions as everyone else.

Highly educated people (like the readers of *Nature*, one of the world's finest scientific journals) score just as badly on our fact questions as everyone else, and often even worse.

People with extraordinary expertise in one field score just as badly on our fact questions as everyone else.

I had the honor of attending the 64th Lindau Nobel Laureate Meeting, and addressing a large group of talented young scientists and Nobel laureates in physiology and medicine. They were the acknowledged intellectual elite of their field, and yet on the question about child vaccination they scored worse than any public polls: 8 percent got the answer right. (After this I never take it for granted that brilliant experts will know anything about closely related fields outside their specializations.)

Being intelligent—being good with numbers, or being well educated, or even winning a Nobel Prize—is not a shortcut to global factual knowledge. Experts are experts only within their field.

And sometimes "experts" are not experts even in their own fields. Many activists present themselves as experts. I have presented at all kinds of activist conferences because I believe educated activists can be absolutely crucial for improving the world. Recently I presented at a conference on women's rights. I strongly support their cause. Two hundred ninety-two brave young feminists had traveled to Stockholm from across the world to coordinate their struggle to improve women's access to education. But only 8 percent knew that 30-year-old women have spent on average only one year less in school than 30-year-old men.

I am absolutely not saying that everything is OK with girls' education. On Level 1, and especially in a small number of countries, many girls still do not go to primary school, and there are huge problems with girls' and women's access to secondary and higher education. But in fact, on Levels 2, 3, and 4, where 6 billion people live, girls are going to school as much as, or more than, boys. This is something amazing! It is something that activists for women's education should know and celebrate.

I could have picked other examples. This is not about activists for women's rights, in particular. Almost every activist I have ever met, whether deliberately or, more likely, unknowingly, exaggerates the problem to which they have dedicated themselves.

FACT QUESTION 11

In 1996, tigers, giant pandas, and black rhinos were all listed as endangered. How many of these three species are more critically endangered today?

- ☐ A: Two of them
- ☐ B: One of them
- ☐ C: None of them

FACT QUESTION 11 RESULTS: percentage who answered correctly.

Tigers, giant pandas, and black rhinos were listed as threatened species in 1996. Since then, have any of these become more critically endangered? (Correct answer: none of them.)

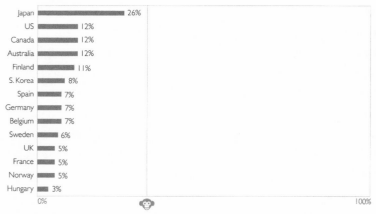

Sources: Ipsos MORI[1] & Novus[1]

Humans have plundered natural resources across the planet. Natural habitats have been destroyed and many animals hunted to extinction. This is clear. But activists who devote themselves to protecting vulnerable animals and their habitats tend to make the same mistake I've just described: desperately trying to make people care, they forget about progress.

A serious problem requires a serious database. I strongly recommend visiting the Red List, where you can access the status of all endangered species in the world, as updated by a global community of high-quality researchers who track the wild populations of different animals and collaborate to monitor the trend. Guess what? If I check the Red List or World Wildlife Fund (WWF) today, I can see how, despite declines in some local populations and some subspecies, the total wild populations of tigers, giant pandas, and black rhinos have all increased over the past years. It was worth paying for all those panda stickers on the doors all around Stockholm. Yet only 6 percent of the Swedish public knows that their support has had any effect.

There has been progress in human rights, animal protection, women's education, climate awareness, catastrophe relief, and many other areas where activists raise awareness by saying that things are getting worse. That progress is often largely thanks to these activists. Maybe they could achieve even more, though, if they did not have such a singular perspective—if they had a better understanding themselves of the progress that had been made, and a greater willingness to communicate it to those they seek to engage. It can be energizing to hear evidence of progress rather than a constant restatement of the problem. UNICEF, Save the Children, Amnesty, and the human rights and environmental movements miss this opportunity again and again.

Hammers and Nails

You probably know the saying "give a child a hammer and everything looks like a nail."

When you have valuable expertise, you like to see it put to use. Sometimes an expert will look around for ways in which their hard-won knowledge and skills can be applicable beyond where it's actually useful. So, people with math skills can get fixated on the numbers. Climate activists argue for solar everywhere. And physicians promote medical treatment where prevention would be better.

Great knowledge can interfere with an expert's ability to see what actually works. All these solutions are great for solving some problem, but none of them will solve all problems. It is better to look at the world in lots of different ways.

Numbers Are Not the Single Solution

I don't love numbers. I am a huge, *huge* fan of data, but I don't love it. It has its limits. I love data only when it helps me to understand the reality behind the numbers, i.e., people's lives. In my research, I have needed the data to test my hypotheses, but the hypotheses themselves often emerged from talking to, listening to, and observing people. Though we absolutely need numbers to understand the world, we should be highly skeptical about conclusions derived purely from number crunching.

The prime minister of Mozambique from 1994 to 2004, Pascoal Mocumbi, visited Stockholm in 2002 and told me that his country was making great economic progress. I asked him how he knew; after all, the quality of the economic statistics in Mozambique was probably not very good. Had he looked at GDP per capita?

"I do look at those figures," he said. "but they are not so accurate. So I have also made it a habit to watch the marches on May first every year. They are a popular tradition in our country. And I look at people's feet, and what kind of shoes they have. I know that people do their best to look good on that day. I know that they cannot borrow their friend's shoes, because their friend will be out marching too. So I look. And I can see if they walk barefoot, or if they have bad shoes, or if they have good shoes. And I can compare what I see with what I saw last year.

"Also, when I travel across the country, I look at the construction going on. If the grass is growing over new foundations, that is bad. But if they keep putting new bricks on, then I know people have money to invest, not just to consume day to day."

A wise prime minister looks at the numbers, but not *only* at the numbers.

And of course some of the most valued and important aspects of human development cannot be measured in numbers at all. We can estimate suffering from disease using numbers. We can measure improvements in material living conditions using numbers. But the end goal of economic growth is individual freedom and culture, and these values are difficult to capture with numbers. The idea of measuring human progress in numbers seems completely bizarre to many people. I often agree. The numbers will never tell the full story of what life on Earth is all about.

The world cannot be understood without numbers. But the world cannot be understood with numbers alone.

Medicine Is Not the Single Solution

Medical professionals can become very single-minded about medicine, or even a particular kind of medicine.

In the 1950s, a Danish public health doctor, Halfdan Mahler, suggested to the World Health Organization a way to eradicate tuberculosis. His project sent small buses with X-ray machines trundling around villages in India. It was a simple idea: eradicate one disease, and it's gone. The plan was to X-ray the whole population, find those with TB, and treat them. But it failed because the people got angry. They all had tons of urgent health problems, and finally here was a bus with nurses and doctors. But instead of fixing a broken bone, or giving fluids for diarrhea, or helping a woman in childbirth, they wanted to X-ray everyone for a disease they had never heard of.

Out of the failure of this attempt to eradicate one single disease came the insight that, instead of fighting this disease or that disease, it is wiser to provide and gradually improve primary health care for all.

In another part of the medical world, the profits of Big Pharma companies have been dropping. Most of them are fixated on developing a new, revolutionary, life-prolonging medicine. I try to persuade them that the next big boost in world life expectancy (and their profits) will probably come not from a pharmacological breakthrough but from a business model breakthrough. Big Pharma is currently failing to reach huge markets in countries on Levels 2 and 3, where hundreds of millions of people, like the diabetes patient we met in Kerala, need drugs that have already been discovered, but at more reasonable prices. If the pharmaceutical companies were better at adjusting their prices for different countries and different customers, they could make their next fortune with what they already have.

Experts in maternal mortality who understand the point about hammers and nails can see that the most valuable intervention for saving the lives of the poorest mothers is not training more local nurses to perform C-sections, or better treatment of severe bleeding or infections, but the availability of transport to the local hospital. The hospitals were of limited use if women could not reach them: if there were no ambulances, or no roads for the ambulances to travel on. Similarly, educators know that it is often the availability of electricity rather than more textbooks or even more teachers in the classroom that has the most impact on learning, as students can do their homework after sunset.

Where Gynecologists Never Put Their Fingers

I was talking to some gynecologists whose job it was to collect data about sexually transmitted diseases in poor communities. These professionals were ready to put their fingers anywhere on people, and to ask them all kinds of questions about their sexual activities. I was interested to know whether some STDs were more common in some income groups, and so I asked them to include a question about income on their forms. They looked at me and said, "What? You can't ask people about their incomes. That is an extremely private question." The one place they didn't want to put their fingers was in people's wallets.

Some years later, I met the team at the World Bank who organized the global income surveys and I asked them to include questions about sexual activity in their survey. I was still wondering about any relationships between sexual behavior and income levels. Their reaction was more or less the same. They were happy to ask people all kinds of questions about their income, the black market, and so on. But sex? Absolutely not.

It's strange where people end up drawing their lines and how well behaved they feel if they stay inside their boxes.

The Ideologues

A big idea can unite people like nothing else and allow us to build the society of our dreams. Ideology has given us liberal democracy and public health insurance.

But ideologues can become just as fixated as experts and activists on their one idea or one solution, with even more harmful outcomes.

The absurd consequences of focusing fanatically on a single idea, like free markets or equality, instead of on measuring performance and doing what works are obvious to anyone who spends much time look-ing at the realities of life in Cuba and the United States.

Cuba: The Healthiest of the Poor

I spent some time in Cuba in 1993, investigating a devastating epidemic that was affecting 40,000 people. I had several encounters with President Fidel Castro himself, and I met many skilled, highly educated, and dedi-cated professionals at the Ministry of Health doing their best within an inflexible and oppressive system. Having lived and worked in a commu-nist country (Mozambique), I went to Cuba with great curiosity but no romantic ideas whatsoever, and I didn't develop any while I was there.

I could tell you countless stories of the nonsense I saw in Cuba: the local moonshine, a toxic fluorescent concoction brewed inside TV tubes using water, sugar, and babies' poopy diapers to provide the yeast required for fermentation; the hotels that hadn't planned for any guests and so had no food, a problem we solved by driving to an old people's home and eating their leftovers from the standard adult food rations; my Cuban colleague who knew his children would be ex-pelled from university if he sent a Christmas card to his cousin in Miami; the fact that I had to explain my research methods to Fidel Castro personally to get approval. I will restrain myself and just tell you why I was there and what I discovered.

In late 1991, the poor farmers in the tobacco-growing province of Pinar del Río had started to go color blind and then experience neurological problems with a loss of feeling in the arms and legs. Cuban epidemiologists had investigated and were now seeking outside help. Since the Soviet Union had just collapsed, no help could come from that direction, and in searching the literature for the few research-ers in the world with experience of neurological pandemics among poor farmers, they hit on me. Conchita Huergo, a member of the Cuban politburo, met me at the airport, and on my first day Fidel himself appeared, accompanied by armed guards, to check me over. His black sneakers squeaked on the cement floor as he circled round me.

I spent three months investigating. I concluded that the poor farmers were suffering not from a mass poisoning from black market food (as rumor had it), nor from some germ causing metabolic problems, but from simple nutritional deficiency caused by global macroeconomics. The Soviet boats that had until recently been arriving full of potatoes and leaving full of Cuban sugar and cigars had not come this year. All food was strictly rationed. The people had given what little nutritious food they had to the children, the pregnant, and the old, and the heroic adults had eaten only rice and sugar. I presented this all as carefully as I could because the clear implication was that government planning had failed to provide enough food for its people. The planned economy had failed. I was thanked and sent home.

One year later I was invited back to Havana to give a presentation to the Ministry of Health on "Health in Cuba in a Global Perspective." The Cuban government had by this point, with the help of the Venezuelan government, regained the ability to feed the Cuban people.

I showed them Cuba's special position on my health and wealth bubble chart. It had a child survival rate as high as that of the United States, on only one-quarter of the income. The minister of health jumped onstage directly after I had finished and summarized my message. "We Cubans are the healthiest of the poor," he said. There was a big round of applause and that turned out to be the end of the session.

However, that was not the message that everyone had taken from my presentation. As I moved toward the refreshments, a young man gently grabbed my arm. He softly dragged me out of the flow of the crowd, explaining that he worked with health statistics. Then he leaned his head close to mine and with his mouth close to my ear he courageously whispered, "Your data is correct but the conclusion of the minister is completely wrong." He looked at me as if it were a quiz, then answered his own question. "We are not the healthiest of the poor, we are the poorest of the healthy."

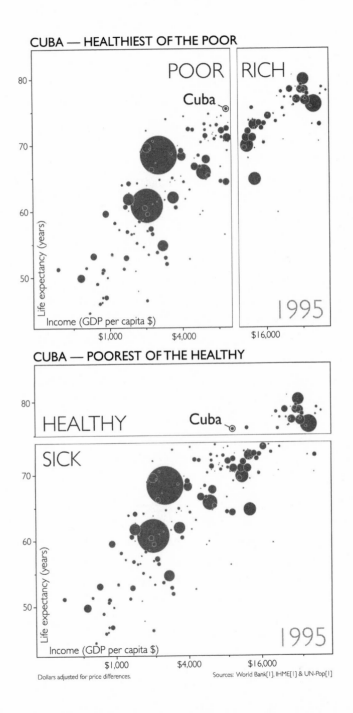

He let go of my arm and swiftly walked away, smiling. Of course, he was right. The Cuban minister had described things from the government's single-minded perspective, but there was also another way of looking at things. Why be pleased with being the healthiest of the poor? Don't the Cuban people deserve to be as rich, and as free, as those in other healthy states?

The United States: The Sickest of the Rich

Which brings us to the United States. Just as Cuba is the poorest of the healthy because of its commitment to a single idea, the United States is the sickest of the rich.

Ideologues will invite you to contrast the United States with Cuba. They will insist you must be for one or the other. If you would prefer to live in the United States than in Cuba, they say, then you must reject everything the government does in Cuba, and you must support what Cuba's government rejects—the free market. To be clear, I would definitely prefer to live in the United States than in Cuba, but I don't find it helpful to think like this. It is single-minded and very misleading. If it is being ambitious, the United States should seek to compare itself not to Cuba, a communist country on Level 3, but to other capitalist countries on Level 4. If US politicians want to make fact-based decisions, they should be driven not by ideology but by the numbers. And if I were to choose where to live, I would choose based not on ideology but on what a country delivers to its people.

The United States spends more than twice as much per capita on health care as other capitalist countries on Level 4—around $9,400 compared to around $3,600—and for that money its citizens can expect lives that are three years shorter. The United States spends more per capita on health care than any other country in the world, but 39 countries have longer life expectancies.

39 COUNTRIES BEAT US LIFE SPAN
but nobody beats the US on health spending

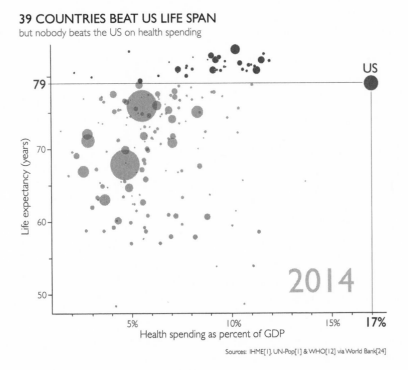

Sources: IHME[1], UN-Pop[1] & WHO[12] via World Bank[24]

Instead of comparing themselves with extreme socialist regimes, US citizens should be asking why they cannot achieve the same levels of health, for the same cost, as other capitalist countries that have similar resources. The answer is not difficult, by the way: it is the absence of the basic public health insurance that citizens of most other countries on Level 4 take for granted. Under the current US system, rich, insured patients visit doctors more than they need, running up costs, while poor patients cannot afford even simple, inexpensive treatments and die younger than they should. Doctors spend time that could be used to save lives or treat illness providing unnecessary, meaningless care. What a tragic waste of physician time.

Actually, to be completely accurate I should say that there is a small number of rich countries with life expectancies as low as that in the United States: the rich Gulf states of Oman, Saudi Arabia, Bahrain, United Arab Emirates, and Kuwait. But these states have a very different history. Until the 1960s when they really started getting rich on oil, their populations were poor and illiterate. Their health systems have been built in just two generations. Unlike the United States, these states are not constrained by a suspicion of anything governmental and I would not be surprised if within a couple of years they all had higher life expectancies than the United States. Perhaps the United States will then be less reluctant to learn from them than it is to learn from Western European countries.

The communist system in Cuba is an example of the danger of getting hooked on a single perspective: the seemingly reasonable but actually bizarre idea that a central government can solve all its people's problems. I can understand why people looking at Cuba and its inefficiencies, poverty, and lack of freedom would decide that governments should never be allowed to plan societies.

The health-care system in the United States is also suffering from the single-perspective mind-set: the seemingly reasonable but actually bizarre idea that the market can solve all a nation's problems. I can understand why people looking at the United States and its inequalities and health-care outcomes would decide that private markets and competition should never be allowed anywhere near the delivery of public goods.

As with most discussions about the private versus the public sector, the answer is not either/or. It is case-by-case, and it is both. The challenge is to find the right balance between regulation and freedom.

Even Democracy Is Not the Single Solution

This is risky, but I am going to argue it anyway. I strongly believe that liberal democracy is the best way to run a country. People like me, who believe this, are often tempted to argue that democracy leads to, or is even a requirement for, other good things, like peace, social progress, health improvements, and economic growth. But here's the thing, and it is hard to accept: the evidence does not support this stance.

Most countries that make great economic and social progress are not democracies. South Korea moved from Level 1 to Level 3 faster than any country had ever done (without finding oil), all the time as a military dictatorship. Of the ten countries with the fastest economic growth in the years 2012–2016, nine of them score low on democracy.

Anyone who claims that democracy is a necessity for economic growth and health improvements will risk getting contradicted by reality. It's better to argue for democracy as a goal in itself instead of as a superior means to other goals we like.

There is no single measure—not GDP per capita, not child mortality (as in Cuba), not individual freedom (as in the United States), not even democracy—whose improvement will guarantee improvements in all the others. There is no single indicator through which we can measure the progress of a nation. Reality is just more complicated than that.

The world cannot be understood without numbers, nor through numbers alone. A country cannot function without a government, but the government cannot solve every problem. Neither the public sector nor the private sector is always the answer. No single measure of a good society can drive every other aspect of its development. It's not either/or. It's both and it's case-by-case.

Factfulness

Factfulness is . . . recognizing that a single perspective can limit your imagination, and remembering that it is better to look at problems from many angles to get a more accurate understanding and find practical solutions.

To control the single perspective instinct, **get a toolbox, not a hammer.**

- **Test your ideas.** Don't only collect examples that show how excellent your favorite ideas are. Have people who disagree with you test your ideas and find their weaknesses.

- **Limited expertise.** Don't claim expertise beyond your field: be humble about what you don't know. Be aware too of the limits of the expertise of others.

- **Hammers and nails.** If you are good with a tool, you may want to use it too often. If you have analyzed a problem in depth, you can end up exaggerating the importance of that problem or of your solution. Remember that no one tool is good for everything. If your favorite idea is a hammer, look for colleagues with screwdrivers, wrenches, and tape measures. Be open to ideas from other fields.

- **Numbers, but not *only* numbers.** The world cannot be understood without numbers, and it cannot be understood with numbers alone. Love numbers for what they tell you about real lives.

- **Beware of simple ideas and simple solutions.** History is full of visionaries who used simple utopian visions to justify terrible actions. Welcome complexity. Combine ideas. Compromise. Solve problems on a case-by-case basis.

CHAPTER NINE

THE BLAME INSTINCT

About magic washing machines and money-making robots

Let's Beat Up Grandma

I was lecturing at Karolinska Institutet, explaining that the big pharmaceutical companies do hardly any research on malaria and nothing at all on sleeping sickness or other illnesses that affect only the poorest.

A student sitting in the front said, "Let's punch them in the face."

"Aha," I said. "I am actually going to Novartis in the fall." (Novartis is a global pharma company based in Switzerland, and I had been invited to give a lecture there.) "If you explain to me what I will achieve and who I should punch, I could try it. Who should I punch in the face? Is it just anybody who works there?"

"No, no, no, no. It's the boss," said that guy.

"Aha. OK. It is Daniel Vasella." That was the name of the boss back then. "Well, I do know Daniel Vasella a bit. When I see him in the fall, should I punch him in the face? Will everything be good then? Will he become a good boss and realize that he should change the company's research priorities?"

A student farther back answered, "No, you have to punch the board members in the face."

"Well, that is actually interesting because I will probably speak in front of the board in the afternoon. So then I'll stay calm in the morning when I see Daniel, but when I get to the boardroom I'll walk around and punch as many as I can. I will of course not have time to knock everyone down . . . I have no fighting experience and there is security there, so they will probably stop me after three or four. But should I do that, then? You think this will make the board change its research policy?"

"No," said a third student. "Novartis is a public company. It's not the boss or the board who decides. It's the shareholders. If the board changes its priorities the shareholders will just elect a new board."

"That's right," I said. "It's the shareholders who want this company to spend their money on researching rich people's illnesses. That's how they get a good return on their shares."

So there's nothing wrong with the employees, the boss, or the board, then.

"Now, the question is"—I looked at the student who had first suggested the face punching—"who owns the shares in these big pharmaceutical companies?"

"Well, it's the rich." He shrugged.

"No. It's actually interesting because pharmaceutical shares are very stable. When the stock market goes up and down, or oil prices go up and down, pharma shares keep giving a pretty steady return. Many other kinds of companies' shares follow the economy—they do better or worse as people go on spending sprees or cut back—but the cancer patients always need treatment. So who owns the shares in these stable companies?"

My young audience looked back at me, their faces like one big question mark.

"It's retirement funds."

Silence.

"So maybe I don't have to do any punching, because I will not meet the shareholders. But you will. This weekend, go visit your grandma and punch her in the face. If you feel you need someone to blame and punish, it's the seniors and their greedy need for stable stocks.

"And remember last summer, when you went backpacking and grandma gave you a little extra travel money? Well. Maybe you should give that back to her, so she can give it back to Novartis and ask them to invest in poor people's health. Or maybe you spent it already, and you should punch yourself in the face."

The Blame Instinct

The blame instinct is the instinct to find a clear, simple reason for why something bad has happened. I had this instinct just recently when I was taking a shower in a hotel and turned the warm handle up to maximum. Nothing happened. Then, seconds later, I was being burned by scorching water. In those moments, I was furious with the plumber, and then the hotel manager, and then the person who might be running cold water next door. But no one was to blame. No one had intentionally caused me harm or been neglectful, except perhaps me, when I didn't have the patience to turn the warm handle more gradually.

It seems that it comes very naturally for us to decide that when things go wrong, it must be because of some bad individual with bad intentions. We like to believe that things happen because someone wanted them to, that individuals have power and agency: otherwise, the world feels unpredictable, confusing, and frightening.

The blame instinct makes us exaggerate the importance of individuals or of particular groups. This instinct to find a guilty party derails our ability to develop a true, fact-based understanding of the world: it steals our focus as we obsess about someone to blame, then blocks our learning because once we have decided who to punch in the face we stop looking for explanations elsewhere. This undermines our ability to solve the problem, or prevent it from happening again, because we are stuck with oversimplistic finger pointing, which distracts us from the more complex truth and prevents us from focusing our energy in the right places.

For example, blaming an airplane crash on a sleepy pilot will not help to stop future crashes. To do that, we must ask: Why was he sleepy? How can we regulate against sleepy pilots in the future? If we stop thinking when we find the sleepy pilot, we make no progress. To understand most of the world's significant problems we have to look beyond a guilty individual and to the system.

The same instinct is triggered when things go well. "Claim" comes just as easily as "blame." When something goes well, we are very quick to give the credit to an individual or a simple cause, when again it is usually more complicated.

If you really want to change the world you have to understand it. Following your blame instinct isn't going to help.

Playing the Blame Game

The blame game often reveals our preferences. We tend to look for bad guys who confirm our existing beliefs. Let's look at some of the people we most love to point the finger at: evil businessmen, lying journalists, and foreigners.

Business

I always try to be analytical, but even so, I am often floored by my instincts. This particular time, perhaps I had been reading too many cartoons featuring Scrooge McDuck, Donald Duck's rich, greedy uncle. Perhaps back then I was as lazy in my thinking about commercial pharma as my students were many years later. At any rate, when UNICEF asked me to investigate a bid for a contract to provide malaria tablets to Angola, I got suspicious. The numbers looked odd and all I could think was that I was going to uncover a scam. Some dishonest business was trying to rip off UNICEF and I was going to find out how.

UNICEF runs competitive bids for pharmaceutical companies to provide it with medicines over a ten-year period. The length and size of the contracts make them attractive and bidders tend to offer very good prices. However, on this occasion, a small family business called Rivopharm, based in Lugano in the Swiss Alps, had put in an unbelievably low bid: in fact, the price they wanted per pill was lower than the cost of the raw materials.

My job was to go over there and find out what was going on. I flew to Zürich, then took a small plane to the little airport in Lugano. I was expecting to be met by a representative of a shabby, corner-cutting outfit but was instead whisked away in a limousine and deposited at the most luxurious hotel I had ever been in. I rang home to Agneta and whispered to her, "Silk sheets."

The next morning I was driven out to the factory to inspect. I shook the manager's hand then went straight in with my questions: "You buy the raw material from Budapest, turn it into pills, put the pills in containers, put the containers in boxes, put the boxes in a shipping container, and get the container to Genoa. How can you do all of that for less than the cost of the raw materials? Do you get some special price from the Hungarians?"

"We pay the same price as everyone else to the Hungarians," he told me.

"And you pick me up in a limousine? Where are you making your money?"

He smiled. "It works like this. A few years ago we saw that robotics was going to change this industry. We built this little factory, with the world's fastest pill-making machine, which we invented. All our other processes are highly automated too. The big companies' factories look like craftsmen's workshops compared with us. So, we order supplies from Budapest. On Monday at six a.m. the active ingredient chloroquine arrives here on the train. By Wednesday afternoon, a year's supply of malaria pills for Angola are packed in boxes ready to ship. By Thursday morning they are at the port in Genoa. UNICEF's buyer inspects the pills and signs that he received them, and the money is paid that day into our Zürich bank account."

"But come on. You are selling it for less than you bought it for."

"That's right. The Hungarians give us 30 days' credit and UNICEF pays us after only four of those days. That gives us 26 days left to earn interest while the money is sitting in our account."

Wow. I couldn't find words. I hadn't even thought of that option.

My mind had been blocked with the idea that UNICEF were the good guys and pharma were the bad guys with an evil plot. I had been completely ignorant about the innovative power of small businesses. They turned out to be good guys too, with a fantastic ability to find cheaper solutions.

Journalists

It is fashionable for intellectuals and politicians to point a finger at the media and blame them for not reporting the truth. Maybe it even seemed like I was doing that myself in earlier chapters.

Instead of pointing our fingers at journalists, we should be asking: Why does the media present such a distorted picture of the world? Do journalists really mean to give us a distorted picture? Or could there be another explanation?

(I am not getting into the debate about deliberately manufactured fake news. That is something else altogether and nothing to do with journalism. And by the way, I do not believe that fake news is the major culprit for our distorted worldview: we haven't only just started to get the world wrong, I think we have always gotten it wrong.)

In 2013, we posted results from Gapminder's Ignorance Project online. The findings quickly became top stories on both BBC and CNN. The two channels posted our questions on their websites so people could test themselves and they got thousands of comments trying to analyze why the heck people were getting such worse-than-random bad results.

One comment caught our attention: "I bet no member of the media passed the test."

We got excited by this idea and decided to try to test it, but the polling companies said it was impossible to get access to groups of journalists. Their employers refused to let them be tested. Of course, I understood. No one likes their authority to be questioned and it would be very embarrassing for a serious news outlet to be shown to be employing journalists who knew no more than chimpanzees.

When people tell me things are impossible, that's when I get really excited to try. In my calendar for that year were two media conferences, so I took our polling devices along. A 20-minute lecture is too short for all my questions, but I could ask some. Here are the results. I also include in the table the results from a conference of leading documentary film producers—people from the BBC, PBS, National Geographic, the Discovery Channel, and so on.

JOURNALISTS AND FILMMAKERS DO NOT BEAT THE CHIMPS

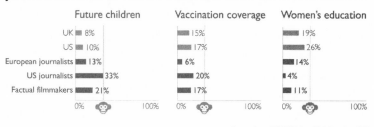

Sources: Ipsos MORI[1], Novus[1] & Gapminder[27]

It seems that these journalists and filmmakers know no more than the general public, i.e., less than chimpanzees.

If this is the case for journalists and documentarians in general—and I have no reason to believe knowledge levels would be higher among other groups of reporters, or that they would have done better with other questions—then they are not guilty. Journalists and documentarians are not lying—i.e., not deliberately misleading us—when they produce dramatic reports of a divided world, or of "nature striking back," or of a population crisis, discussed in serious tones with wistful piano music in the background. They do not necessarily have bad intentions, and blaming them is pointless. Because most of the journalists and filmmakers who inform us about the world are themselves misled. Do not demonize journalists: they have the same mega misconceptions as everyone else.

Our press may be free, and professional, and truth-seeking, but independent is not the same as representative: even if every report is itself completely true, we can still get a misleading picture through the sum of true stories reporters choose to tell. The media is not and cannot be neutral, and we shouldn't expect it to be.

The journalists' poll results are pretty disastrous. They are the knowledge equivalent of a plane crash. But it is no more helpful to blame the journalists than it is to blame a sleepy pilot. Instead, we have to seek to understand why journalists have a distorted worldview (answer: because they are human beings, with dramatic instincts) and what systemic factors encourage them to produce skewed and overdramatic news (at least part of the answer: they must compete for their consumers' attention or lose their jobs).

When we understand this we will realize that it is completely unrealistic and unfair to call for the media to change in this way or that so that it can provide us with a better reflection of reality. Reflecting reality is not something the media can be expected to do. You should not expect the media to provide you with a fact-based worldview any more than you would think it reasonable to use a set of holiday snaps of Berlin as your GPS system to help you navigate around the city.

Refugees

In 2015, 4,000 refugees drowned in the Mediterranean Sea as they tried to reach Europe in inflatable boats. Images of children's bodies washed up on the shores of holiday destinations evoked horror and compassion. What a tragedy. In our comfortable lives on Level 4 in Europe and elsewhere, we started thinking: How could such a thing happen? Who was to blame?

We soon worked it out. The villains were the cruel and greedy smugglers who tricked desperate families into handing over 1,000 euros per person for their places in inflatable death traps. We stopped thinking and comforted ourselves with images of European rescue boats saving people from the wild waters.

But why weren't the refugees traveling to Europe on comfortable planes or ferry boats instead of traveling over land to Libya or Turkey and then entrusting their lives to these rickety rubber rafts? After all, all EU member states were signed up to the Geneva Convention, and it was clear that refugees from war-torn Syria would be entitled to claim asylum under its terms. I started to ask this question of journalists, friends, and people involved in the reception of the asylum seekers, but even the wisest and kindest among them came up with very strange answers.

Perhaps they could not afford to fly? But we knew that the refugees were paying 1,000 euros for each place on a rubber dinghy. I went online and checked and there were plenty of tickets from Turkey to Sweden or from Libya to London for under 50 euros.

Maybe they couldn't reach the airport? Not true. Many of them were already in Turkey or Lebanon and could easily get to the airport. And they can afford a ticket, and the planes are not overbooked. But at the check-in counter, they are stopped by the airline staff from getting onto the plane. Why? Because of a European Council Directive from 2001 that tells member states how to combat illegal immigration. This directive says that every airline or ferry company that brings a person without proper documents into Europe must pay all the costs of returning that person to their country of origin. Of course the directive also says that it doesn't apply to refugees who want to come to Europe based on their rights to asylum under the Geneva Convention, only to illegal immigrants. But that claim is meaningless. Because how should someone at the check-in desk at an airline be able to work out in 45 seconds whether someone is a refugee or is not a refugee according to the Geneva Convention? Something that would take the embassy at least eight months? It is impossible. So the practical effect of the reasonable-sounding directive is that commercial airlines will not let anyone board without a visa. And getting a visa is nearly impossible because the European embassies in Turkey and Libya do not have the resources to process the applications. Refugees from Syria, with the theoretical right to enter Europe under the Geneva Convention, are therefore in practice completely unable to travel by air and so must come over the sea.

Why, then, must they come in such terrible boats? Actually, EU policy is behind that as well, because it is EU policy to confiscate the boats when they arrive. So boats can be used for one trip only. The smugglers could not afford to send the refugees in safe boats, like the fishing boats that brought 7,220 Jewish refugees from Denmark to Sweden over a few days in 1943, even if they wanted to.

Our European governments claim to be honoring the Geneva Convention that entitles a refugee from a severely war-torn country to apply for and receive asylum. But their immigration policies make a mockery of this claim in practice and directly create the transport market in which the smugglers operate. There is nothing secret about this; in fact it takes some pretty blurry or blocked thinking not to see it.

We have an instinct to find someone to blame, but we rarely look in the mirror. I think smart and kind people often fail to reach the terrible, guilt-inducing conclusion that our own immigration policies are responsible for the drownings of refugees.

Foreigners

Remember the Indian official in chapter 5 who so persuasively rejected the claim that India and China should be taking the blame for climate change? I used the story then to talk about the importance of per-person measures, but of course it is also about how finding someone to blame can distract us from looking at the whole system.

The idea that India, China, and other countries moving up the levels should be blamed for climate change, and that their populations should be forced to live poorer lives in order to address it, is shockingly well established in the West. I remember, during a lecture about global trends at Tech University in Vancouver, an outspoken student saying with despair in her voice, "They can't live like us. We can't let them continue developing like this. Their emissions will kill the planet." It is shocking how often I hear Westerners talking as if they hold remote controls in their hands and can make decisions about billions of lives elsewhere, just by pressing a button. Looking around, I realized that her fellow students were not reacting at all. They agreed with her.

Most of the human-emitted CO_2 accumulated in the atmosphere was emitted over the last 50 years by countries that are now on Level 4. Canada's per capita CO_2 emissions are still twice as high as China's and eight times as high India's. In fact, do you know how much of all the fossil fuel burned each year is burned by the richest billion? More than half of it. Then the second-richest billion burns half of what's left, and so on and so on, down to the poorest billion, who are responsible for only 1 percent.

CO₂ EMISSIONS BY INCOME

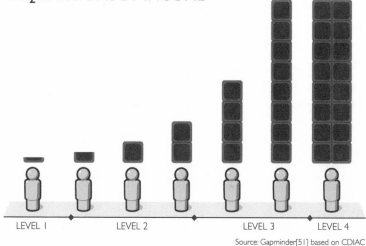

Source: Gapminder[51] based on CDIAC

It will take at least two decades for the poorest billion to struggle from Level 1 to Level 2—increasing their contribution to global CO_2 emissions by roughly 2 percent. It will take several decades more for them to get up to Levels 3 and 4.

In these circumstances, it is a testament to the blame instinct how easily we in the West seem to shift responsibility away from ourselves and onto others. We say that "they" cannot live like us. The right thing to say is, "We cannot live like us."

The Foreign Disease

The body's largest organ is the skin. Before modern medicine, one of the worst imaginable skin diseases was syphilis, which would start as itchy boils and then eat its way into the bones until it exposed the skeleton. The microbe that caused this disgusting sight and unbearable pain had different names in different places. In Russia it was called the Polish disease. In Poland it was the German disease; in Germany, the French disease; and in France, the Italian disease. The Italians blamed back, calling it the French disease.

The instinct to find a scapegoat is so core to human nature that it's hard to imagine the Swedish people calling the open sores the Swedish disease, or the Russians calling it the Russian disease. That's not how people work. We need someone to blame and if a single foreigner came here with the disease, then we would happily blame a whole country. No further investigation needed.

Blame and Claim

The blame instinct drives us to attribute more power and influence to individuals than they deserve, for bad or good. Political leaders and CEOs in particular often claim they are more powerful than they are.

Powerful Leaders?

For example, Mao was undoubtedly an extraordinarily powerful figure whose actions had direct consequences for 1 billion people. Most often when I show the low birth numbers in Asia, someone says, "That must be because of Mao's one-child policy."

But the infamous one-child policy had less influence on birth rates than is commonly thought. The huge, fast drop from six to three babies per woman in China happened in the ten years preceding the one-child policy. During the 36 years the policy was in place, the number never fell below 1.5, though it did in many other countries without enforcement, like Ukraine, Thailand, and South Korea. In Hong Kong, where again the one-child policy didn't apply, the number dropped even below one baby per woman. All this suggests that there were other factors at play here than the decisive command of a powerful man. And it wasn't even Mao's policy. It was introduced after his death.

The pope is also credited with enormous influence over the sexual behavior of the 1 billion Catholics in the world. However, despite the clear condemnation of the use of contraception by several successive popes, the statistics show that contraceptive use is 60 percent in Catholic-majority countries, compared with 58 percent in the rest of the world. In other words, it is the same. The pope is one of the world's most prominent moral leaders, but it seems that even leaders with great political power or moral authority do not have remote controls that can reach into the bedroom.

The Inside of Sister Linda's Door

In the poorest rural parts of Africa, it is still the nuns who maintain many basic health services. Some of these clever, hardworking, and pragmatic women became my closest colleagues.

Sister Linda, whom I worked with in Tanzania, was a devout Catholic nun who dressed all in black and prayed three times a day. The door to her office was always open—she closed it only during health-care consultations—and on its outside, the first thing you saw as you entered, was a glossy poster of the pope. One day, she and I were in her office and started discussing a sensitive matter. Sister Linda stood up and closed the door, and for the first time I saw what was on its inside: another large poster and, attached to it, hundreds of little bags of condoms. When Sister Linda turned back around and saw my surprised face she smiled—as she often did when discovering my countless stereotypes of women like her. "The families need them to stop both AIDS and babies," she said simply. And then she continued our discussion.

The situation with abortion is different. Mao's one-child policy did have an impact. It resulted in an unknown number of forced abortions and forced sterilizations. Across the world today, women and girls are still being made the victims of religious condemnation of abortion. When abortion is made illegal it doesn't stop abortions from happening, but it does make abortions more dangerous and increase the risk of women dying as a result.

More Likely Suspects

I have argued above that we should look at the systems instead of looking for someone to blame when things go wrong. We should also give more credit to two kinds of systems when things go right. The invisible actors behind most human success are prosaic and dull compared to great, all-powerful leaders. Nevertheless I want to praise them, so let's throw a parade for the unsung heroes of global development: institutions and technology.

Institutions

Only in a few countries, with exceptionally destructive leaders and conflicts, has social and economic development been halted. Everywhere else, even with the most incapable presidents imaginable, there has been progress. It must make one ask if the leaders are that important. And the answer, probably, is no. It's the people, the many, who build a society.

Sometimes, when I turn the water on to wash my face in the morning and warm water comes out just like magic, I silently praise those who made it possible: the plumbers. When I'm in that mode I'm often overwhelmed by the number of opportunities I have to feel grateful to civil servants, nurses, teachers, lawyers, police officers, firefighters, electricians, accountants, and receptionists. These are the people building societies. These are the invisible people working in a web of related services that make up society's institutions. These are the people we should celebrate when things are going well.

In 2014, I went to Liberia to help fight Ebola because I was afraid that if it weren't stopped, it could easily spread to the rest of the world and kill a billion people, causing more harm than any known pandemic in world history. The fight against the lethal Ebola virus was won not by an individual heroic leader, or even by one heroic organization like Médecins Sans Frontières or UNICEF. It was won prosaically and undramatically by government staff and local health workers, who created public health campaigns that changed ancient funeral practices in a matter of days; risked their lives to treat dying patients; and did the cumbersome, dangerous, and delicate work of finding and isolating all the people who had been in contact with them. Brave and patient servants of a functioning society, rarely ever mentioned—but the true saviors of the world.

Technology

The Industrial Revolution saved billions of lives not because it produced better leaders but because it produced things like chemical detergents that could run in automatic washing machines.

I was four years old when I saw my mother load a washing machine for the first time. It was a great day for my mother; she and my father had been saving money for years to be able to buy that machine. Grandma, who had been invited to the inauguration ceremony for the new washing machine, was even more excited. She had been heating water with firewood and hand-washing laundry her whole life. Now she was going to watch electricity do that work. She was so excited that she sat on a chair in front of the machine for the entire washing cycle, mesmerized. To her the machine was a miracle.

It was a miracle for my mother and me too. It was a magic machine. Because that very day my mother said to me, "Now, Hans, we have loaded the laundry. The machine will do the work. So now we can go to the library." In went the laundry, and out came books. Thank you industrialization, thank you steel mill, thank you power station, thank you chemical-processing industry, for giving us the time to read books.

Two billion people today have enough money to use a washing machine and enough time for mothers to read books—because it is almost always the mothers who do the laundry.

FACT QUESTION 12

How many people in the world have some access to electricity?

- ☐ A: 20 percent
- ☐ B: 50 percent
- ☐ C: 80 percent

Electricity is a basic need, which means that the vast majority—almost everyone on Levels 2, 3, and 4—already has it. Still, just one person in four gets the answer right. (The full country breakdown is in the appendix.) The correct answer is the most positive, as usual: 80 percent of people have some access to electricity. It's unstable and there are often power outages, but the world is getting there. One inauguration after another. Home by home.

So let's be realistic about what the 5 billion people in the world who still wash their clothes by hand are hoping for and what they will do everything they can to achieve. Expecting them to voluntarily slow down their economic growth is absolutely unrealistic. They want washing machines, electric lights, decent sewage systems, a fridge to store food, glasses if they have poor eyesight, insulin if they have diabetes, and transport to go on vacation with their families just as much as you and I do.

Unless you are willing to forgo all these things and start hand-washing your jeans and your bedsheets, why should you expect them to? Instead of finding someone to blame and expecting them to take responsibility, what we need in order to save the planet from the huge risks of climate change is a realistic plan. We must put our efforts into inventing new technologies that will enable 11 billion people to live the life that we should expect all of them to strive for. The life we are living now on Level 4, but with smarter solutions.

Who Should You Blame?

It's not the boss or the board or the shareholders who are to blame for the tragic lack of research into the diseases of the poorest. What do we gain from pointing our fingers at them?

Similarly, resist the urge to blame the media for lying to you (mostly they are not) or for giving you a skewed worldview (which mostly they are, but often not deliberately). Resist blaming experts for focusing too much on their own interests and specializations or for getting things wrong (which sometimes they do, but often with good intentions). In fact, resist blaming any one individual or group of individuals for anything. Because the problem is that when we identify the bad guy, we are done thinking. And it's almost always more complicated than that. It's almost always about multiple interacting causes—a system. If you really want to change the world, you have to understand how it actually works and forget about punching anyone in the face.

Factfulness

Factfulness is . . . recognizing when a scapegoat is being used and remembering that blaming an individual often steals the focus from other possible explanations and blocks our ability to prevent similar problems in the future.

To control the blame instinct, **resist finding a scapegoat.**

- **Look for causes, not villains.** When something goes wrong don't look for an individual or a group to blame. Accept that bad things can happen without anyone intending them to. Instead spend your energy on understanding the multiple interacting causes, or system, that created the situation.

- **Look for systems, not heroes.** When someone claims to have caused something good, ask whether the outcome might have happened anyway, even if that individual had done nothing. Give the system some credit.

CHAPTER TEN

THE URGENCY INSTINCT

How "now or never" can block our roads and our minds

Roadblocks and Mental Blocks

"If it's not contagious, then why did you evacuate your children and wife?" asked the mayor of Nacala, eyeing me from a safe distance behind his desk. Out the window, a breathtaking sun was setting over Nacala district and its population of hundreds of thousands of extremely poor people, served by just one doctor—me.

Earlier that day I had arrived back in the city from a poor coastal area in the north named Memba. There I had spent two days using my hands to diagnose hundreds of patients with a terrible, unexplained disease that had completely paralyzed their legs within minutes of onset and, in severe cases, made them blind. And the mayor was right; I wasn't 100 percent sure it was not contagious. I hadn't slept the previous night but had stayed up, poring over my medical textbook, until I had finally concluded that the symptoms I was seeing had not been described before. I'd guessed this was some kind of poison rather than anything infectious, but I couldn't be sure, and I had asked my wife to take our young children and leave the district.

Before I could figure out what to say, the mayor said, "If you think it could be contagious, I must do something. To avoid a catastrophe, I must stop the disease from reaching the city."

The worst-case scenario had already unfolded in the mayor's mind, and immediately spread to mine.

The mayor was a man of action. He stood up and said, "Should I tell the military to set up a roadblock and stop the buses from the north?"

"Yes," I said. "I think it's a good idea. You have to do something."

The mayor disappeared to make some calls.

When the sun rose over Memba the next morning, some 20 women and their youngest children were already up, waiting for the morning bus to take them to the market in Nacala to sell their goods. When they learned the bus had been canceled, they walked down to the beach and asked the fishermen to take them by the sea route instead. The fishermen made room for everyone in their small boats, probably happy to be making the easiest money of their lives as they sailed south along the coast.

Nobody could swim and when the boats capsized in the waves, all the mothers and children and fishermen drowned.

That afternoon I headed north again, past the roadblock, to continue to investigate the strange disease. As I drove through Memba I came across a group of people lining up on the roadside dead bodies they had pulled put of the sea. I ran down to the beach but it was too late. I asked a man carrying the body of a young boy, "Why were all these children and mothers out in those fragile boats?"

"There was no bus this morning," he said. Several minutes later I could still barely understand what I had done. Still today I can't forgive myself. Why did I have to say to the mayor, "You must do something"?

I couldn't blame these tragic deaths on the fishermen. Desperate people who need to get to market of course take the boat when the city authorities for some reason block their road.

I have no way to tell you how I carried on with the work I had to do that day and in the days afterward. And I didn't talk about this to anyone else for 35 years.

But I did carry on with my work and eventually I discovered the cause of the paralytic disease: as I suspected, the people had been poisoned. The surprise was that they had not eaten anything new. The cassava that formed the basis for the local diet had to be processed for three days to make it edible. Everyone had always known that, so no one had ever even heard of anyone who had been poisoned or seen these symptoms. But this year, there had been a terrible harvest across the whole country and the government had been buying processed cassava at the highest price ever. The poor farmers were suddenly able to make that extra money they needed to escape poverty and were selling everything they had. After a successful day of selling, though, they were coming home hungry. So hungry that they couldn't resist eating the unprocessed cassava roots straight from the fields. At 8 p.m. on August 21, 1981, this discovery transformed me from being a district doctor to being a researcher, and I spent the next ten years of my life investigating the interplay among economies, societies, toxins, and food.

Fourteen years later, in 1995, the ministers in Kinshasa, the capital of DR Congo, heard that there was an Ebola outbreak in the city of Kikwit. They got scared. They felt they had to do something. They set up a roadblock.

Again, there were unintended consequences. Feeding the people in the capital became a major problem because the rural area that had always supplied most of their processed cassava was on the other side of the disease-stricken area. The city was hungry and started buying all it could from its second-largest food-producing area. Prices skyrocketed, and guess what? A mysterious outbreak of paralyzed legs and blindness followed.

Nineteen years after that, in 2014, there was an outbreak of Ebola in the rural north of Liberia. Inexperienced people from rich countries got scared and they all came up with the same idea: a roadblock!

At the Ministry of Health, I encountered politicians of a higher quality. They were more experienced, and their experience made them cautious. Their main concern was that roadblocks would destroy the trust of the people abandoned behind them. This would have been absolutely catastrophic: Ebola outbreaks are defeated by contact tracers, who depend on people honestly disclosing everybody they have touched. These heroes were sitting in poor slum dwellings carefully interviewing people who had just lost a family member about every individual their loved one might have infected before dying. Often, of course, the person being interviewed was on that list and potentially infected. Despite the constant fear and wave after wave of rumors, there was no room for drastic, panicky action. The infection path could not be traced with brute force, just patient, calm, meticulous work. One single individual delicately leaving out information about his dead brother's multiple lovers could cost a thousand lives.

When we are afraid and under time pressure and thinking of worst-case scenarios, we tend to make really stupid decisions. Our ability to think analytically can be overwhelmed by an urge to make quick decisions and take immediate action.

Back in Nacala in 1981, I spent several days carefully investigating the disease but less than a minute thinking about the consequences of closing the road. Urgency, fear, and a single-minded focus on the risks of a pandemic shut down my ability to think things through. In the rush to do something, I did something terrible.

The Urgency Instinct

Now or never! Learn Factfulness now! Tomorrow may be too late!

You have reached the final instinct. Now it is time for you to decide. This moment will never come back. Never again will all these instincts be right there at the front of your mind. You have a unique opportunity, today, right now, to capture the insights of this book and completely change the way you think forever. Or you can just finish the book, close it, say to yourself "that was strange," and carry on exactly as before.

But you have to decide now. You have to act now. Will you change the way you think today? Or live in ignorance forever? It's up to you.

You have probably heard something like this before, from a salesperson or an activist. Both use a lot of the same techniques: "Act now, or lose the chance forever." They are deliberately triggering your urgency instinct. The call to action makes you think less critically, decide more quickly, and act now.

Relax. It's almost never true. It's almost never that urgent, and it's almost never an either/or. You can put the book down if you like and do something else. In a week or a month or a year you can pick it up again and remind yourself of its main points, and it won't be too late. That is actually a better way to learn than trying to cram it all in at once.

The urgency instinct makes us want to take immediate action in the face of a perceived imminent danger. It must have served us humans well in the distant past. If we thought there might be a lion in the grass, it wasn't sensible to do too much analysis. Those who stopped and carefully analyzed the probabilities are not our ancestors. We are the offspring of those who decided and acted quickly with insufficient information. Today, we still need the urgency instinct—for example, when a car comes out of nowhere and we need to take evasive action. But now that we have eliminated most immediate dangers and are left with more complex and often more abstract problems, the urgency instinct can also lead us astray when it comes to our understanding the world around us. It makes us stressed, amplifies our other instincts and makes them harder to control, blocks us from thinking analytically, tempts us to make up our minds too fast, and encourages us to take drastic actions that we haven't thought through.

We do not seem to have a similar instinct to act when faced with risks that are far off in the future. In fact, in the face of future risks, we can be pretty slothful. That is why so few people save enough for their retirement.

This attitude toward future risk is a big problem for activists who are working on long timescales. How can they wake us up? How can they galvanize us into action? Very often, it is by convincing us that an uncertain future risk is actually a sure immediate risk, that we have a historic opportunity to solve an important problem and it must be tackled now or never: that is, by triggering the urgency instinct.

This method sure can make us act but it can also create unnecessary stress and poor decisions. It can also drain credibility and trust from their cause. The constant alarms make us numb to real urgency. The activists who present things as more urgent than they are, wanting to call us to action, are boys crying wolf. And we remember how that story ends: with a field full of dead sheep.

Learn to Control the Urgency Instinct. Special Offer! Today Only!

When people tell me we must act now, it makes me hesitate. In most cases, they are just trying to stop me from thinking clearly.

A Convenient Urgency

FACT QUESTION 13

Global climate experts believe that, over the next 100 years, the average temperature will...

☐ A: get warmer
☐ B: remain the same
☐ C: get colder

"We need to create fear!" That's what Al Gore said to me at the start of our first conversation about how to teach climate change. It was 2009 and we were backstage at a TED conference in Los Angeles. Al Gore asked me to help him and use Gapminder's bubble graphs to show a worst-case future impact of a continued increase in CO_2 emissions.

I had a profound respect at that time for Al Gore's achievements in explaining and acting on climate change, and I still do. I am sure you got the fact question at the top of this section right: it's the one question where our audiences always beat the chimps, with the large majority of people (from 94 percent in Finland, Hungary, and Norway, to 81 percent in Canada and the United States, to 76 percent in Japan) knowing very well what drastic change the climate experts are foreseeing. That high level of awareness is in no small part thanks to Al Gore. So is the enormous achievement of the 2015 Paris Agreement on reduction of climate change. He was—and still is—a hero to me. I agreed with him completely that swift action on climate change was needed, and I was excited at the thought of collaborating with him.

But I couldn't agree to what he had asked.

I don't like fear. Fear of war plus the panic of urgency made me see a Russian pilot and blood on the floor. Fear of pandemic plus the panic of urgency made me close the road and cause the drownings of all those mothers, children, and fishermen. Fear plus urgency make for stupid, drastic decisions with unpredictable side effects. Climate change is too important for that. It needs systematic analysis, thought-through decisions, incremental actions, and careful evaluation.

And I don't like exaggeration. Exaggeration undermines the credibility of well-founded data: in this case, data showing that climate change is real, that it is largely caused by greenhouse gases from human activities such as burning fossil fuels, and that taking swift and broad action now would be cheaper than waiting until costly and unacceptable climate change happened. Exaggeration, once discovered, makes people tune out altogether.

I insisted that I would never show the worst-case line without showing the probable and the best-case lines as well. Picking only the worst-case scenario and—worse—continuing the line beyond the scientifically based predictions would fall far outside Gapminder's mission to help people understand the basic facts. It would be using our credibility to make a call to action. Al Gore continued to press his case for fearful animated bubbles beyond the expert forecasts, over several more conversations, until finally I closed the discussion down. "Mr. Vice President. No numbers, no bubbles."

Some aspects of the future are easier to predict than others. Weather forecasts are rarely accurate more than a week into the future. Forecasting a country's economic growth and unemployment rates is also surprisingly difficult. That is because of the complexity of the systems involved. How many things do you need to predict, and how quickly do they change? By next week, there will have been billions of changes of temperature, wind speed, humidity. By next month, billions of dollars will have changed hands billions of times.

In contrast, demographic forecasts are amazingly accurate decades into the future because the systems involved—essentially, births and deaths—are quite simple. Children are born, grow up, have more children, and then die. Each individual cycle takes roughly 70 years.

But the future is always uncertain to some degree. And whenever we talk about the future we should be open and clear about the level of uncertainty involved. We should not pick the most dramatic estimates and show a worst-case scenario as if it were certain. People would find out! We should ideally show a mid-forecast, and also a range of alternative possibilities, from best to worst. If we have to round the numbers we should round to our own disadvantage. This protects our reputations and means we never give people a reason to stop listening.

Insist on the Data

Al Gore's words echoed around my head long after that first conversation.

To be absolutely clear, I am deeply concerned about climate change because I am convinced it is real—as real as Ebola was in 2014. I understand the temptation to raise support by picking the worst projections and denying the huge uncertainties in the numbers. But those who care about climate change should stop scaring people with unlikely scenarios. Most people already know about and acknowledge the problem. Insisting on it is like kicking at an open door. It's time to move on from talking talking talking. Let's instead use that energy to solve the problem by taking action: action driven not by fear and urgency but by data and coolheaded analysis.

So, what is the solution? Well, it's easy. Anyone emitting lots of greenhouse gas must stop doing that as soon as possible. We know who that is: the people on Level 4 who have by far the highest levels of CO_2 emissions, so let's get on with it. And let's make sure we have a serious data set for this serious problem so that we can track our progress.

Looking for the data after my conversation with Al Gore, I was surprised how difficult it was to find. Thanks to great satellite images, we can track the North Pole ice cap on a daily basis. This removes any doubt that it is shrinking from year to year at a worrying speed. So we have good indications of the symptoms of global warming. But when I looked for the data to track the cause of the problem—mainly CO_2 emissions—I found surprisingly little.

The per capita GDP growth of countries on Level 4 was being carefully tracked, with new official numbers released on a quarterly basis. But CO_2 emissions data was being published only once every two years. So I started provoking the Swedish government to do better. In 2009, I started to lobby for quarterly publication of greenhouse gas data: If we cared about it, why weren't we measuring it? How could we claim to be taking this problem seriously if we weren't even tracking our progress?

I am very proud that, since 2014, Sweden now tracks quarterly greenhouse gas emissions (the first and still the only country to do so). This is Factfulness in action. Statisticians from South Korea recently visited Stockholm to learn how they could do the same.

Climate change is way too important a global risk to be ignored or denied, and the vast majority of people living on Level 4 know that. But it is also way too important to be left to sketchy worst-case scenarios and doomsday prophets.

When you are called to action, sometimes the most useful action you can take is to improve the data.

A Convenient Fear

Still, the volume on climate change keeps getting turned up. Many activists, convinced it is the only important global issue, have made it a practice to blame everything on the climate, to make it the single cause of all other global problems.

They grab at the immediate shocking concerns of the day—the war in Syria, ISIS, Ebola, HIV, shark attacks, almost anything you can imagine—to increase the feeling of urgency about the long-term problem. Sometimes the claims are based on strong scientific evidence, but in many cases they are far-fetched, unproven hypotheses. I understand the frustrations of those struggling to make future risks feel concrete in the present. But I cannot agree with their methods.

Most concerning is the attempt to attract people to the cause by inventing the term "climate refugees." My best understanding is that the link between climate change and migration is very, very weak. The concept of climate refugees is mostly a deliberate exaggeration, designed to turn fear of refugees into fear of climate change, and so build a much wider base of public support for lowering CO_2 emissions.

When I say this to climate activists they often tell me that invoking fear and urgency with exaggerated or unsupported claims is justified because it is the only way to get people to act on future risks. They have convinced themselves that the end justifies the means. And I agree that it might work in the short term. *But.*

Crying wolf too many times puts at risk the credibility and reputation of serious climate scientists and the entire movement. With a problem as big as climate change, we cannot let that happen. Exaggerating the role of climate change in wars and conflicts, or poverty, or migration, means that the other major causes of these global problems are ignored, hampering our ability to take action against them. We cannot get into a situation where no one listens anymore. Without trust, we are lost.

And hotheaded claims often entrap the very activists who are using them. The activists defend them as a smart strategy to get people engaged, and then forget that they are exaggerating and become stressed and unable to focus on realistic solutions. People who are serious about climate change must keep two thoughts in their heads at once: they must continue to care about the problem but not become victims of their own frustrated, alarmist messages. They must look at the worst-case scenarios but also remember the uncertainty in the data. In heating up others, they must keep their own brains cool so that they can make good decisions and take sensible actions, and not put their credibility at risk.

Ebola

I described in chapter 3 how, in 2014, I was too slow to understand the dangers of the Ebola outbreak in West Africa. It was only when I saw that the trend line was doubling that I understood. Even in this most urgent and fearful of situations though, I was determined to try to learn from my past mistakes, and act on the data, not on instinct and fear.

The numbers behind the official World Health Organization and the US Centers for Disease Control and Prevention (CDC) "suspected cases" curve were far from certain. Suspected cases means cases that are not confirmed. There were all kinds of issues: for example, people who at some point had been suspected of having Ebola but who, it turned out, had died from some other cause were still counted as suspected cases. As fear of Ebola increased, so did suspicion, and more and more people were "suspected." As the normal health services staggered under the weight of dealing with Ebola and resources had to move away from treating other life-threatening conditions, more and more people were dying from non-Ebola causes. Many of these deaths were also treated as "suspect." So the rising curve of suspected cases got more and more exaggerated and told us less and less about the trend in actual, confirmed cases.

If you can't track progress, you don't know whether your actions are working. So when I arrived at the Ministry of Health in Liberia, I asked how we could get a picture of the number of confirmed cases. I learned within a day that blood samples were being sent to four different labs, and their records, in long and messy Excel spreadsheets, were not being combined. We had hundreds of health-care workers from across the world flying in to take action, and software developers constantly coming up with new, pointless Ebola apps (apps were their hammers and they were desperate for Ebola to be a nail). But no one was tracking whether the action was working or not.

With permission, I sent the four Excel spreadsheets home to Ola in Stockholm, who spent 24 hours cleaning and combining them by hand, and then carrying out the same procedure one more time to make sure the strange thing he saw wasn't a mistake. It wasn't. When a problem seems urgent the first thing to do is not to cry wolf, but to organize the data. To everybody's surprise, the data came back showing that the number of confirmed cases had reached a peak two weeks earlier and was now dropping. The number of suspected cases kept increasing. Meanwhile, in reality, the Liberian people had successfully changed their behavior, eliminating all unnecessary body contact. There was no shaking hands and no hugging. This, and the pedantic obedience to strict hygiene measures being imposed in stores, public buildings, ambulances, clinics, burial sites, and everywhere else was already having the desired effect. The strategy was working, but until the moment Ola sent me the curve, nobody knew. We celebrated and then everybody continued their work, encouraged to try even harder now that they knew what they were doing was actually working.

I sent the falling curve to the World Health Organization and they published it in their next report. But the CDC insisted on sticking to the rising curve of "suspected cases." They felt they had to maintain a sense of urgency among those responsible for sending resources. I understand they were acting from the best of intentions, but it meant that money and other resources were directed at the wrong things. More seriously, it threatened the long-term credibility of epidemiological data. We shouldn't blame them. A long jumper is not allowed to measure her own jumps. A problem-solving organization should not be allowed to decide what data to publish either. The people trying to solve a problem on the ground, who will always want more funds, should not also be the people measuring progress. That can lead to really misleading numbers.

It was data—the data showing that suspected cases were doubling every three weeks—that made me realize how big the Ebola crisis was. It was also data—the data showing that confirmed cases were now falling—that showed me that what was being done to fight it was working. Data was absolutely key. And because it will be key in the future too, when there is another outbreak somewhere, it is crucial to protect its credibility and the credibility of those who produce it. Data must be used to tell the truth, not to call to action, no matter how noble the intentions.

Urgent! Read This Now!

Urgency is one of the worst distorters of our worldview. I know I probably said that about all the other dramatic instincts too, but I think maybe this one really is special. Or perhaps they all come together in this one. The overdramatic worldview in people's heads creates a constant sense of crisis and stress. The urgent "now or never" feelings it creates lead to stress or apathy: "We must do something drastic. Let's not analyze. Let's do something." Or, "It's all hopeless. There's nothing we can do. Time to give up." Either way, we stop thinking, give in to our instincts, and make bad decisions.

The Five Global Risks We *Should* Worry About

I do not deny that there are pressing global risks we need to address. I am not an optimist painting the world in pink. I don't get calm by looking away from problems. The five that concern me most are the risks of global pandemic, financial collapse, world war, climate change, and extreme poverty. Why is it these problems that cause me most concern? Because they are quite likely to happen: the first three have all happened before and the other two are happening now; and because each has the potential to cause mass suffering either directly or indirectly by pausing human progress for many years or decades. If we fail here, nothing else will work. These are mega killers that we must avoid, if at all possible, by acting collaboratively and step-by-step.

(There is a sixth candidate for this list. It is the unknown risk. It is the probability that something we have not yet even thought of will cause terrible suffering and devastation. That is a sobering thought. While it is truly pointless worrying about something unknown that we can do nothing about, we must also stay curious and alert to new risks, so that we can respond to them.)

Global Pandemic

The Spanish flu that spread across the world in the wake of the First World War killed 50 million people—more people than the war had, although that was partly because the populations were already weakened after four years of war. As a result, global life expectancy fell by ten years, from 33 to 23, as you can see from the dip in the curve on page 55. Serious experts on infectious diseases agree that a new nasty kind of flu is still the most dangerous threat to global health. The reason: flu's transmission route. It flies through the air on tiny droplets. A person can enter a subway car and infect everyone in it without them touching each other, or even touching the same spot. An airborne disease like flu, with the ability to spread very fast, constitutes a greater threat to humanity than diseases like Ebola or HIV/AIDS. Protecting ourselves in every possible way from a virus that is highly transmissible and ignores every type of defense is worth the effort, to put it mildly.

The world is more ready to deal with flu than it has been in the past, but people on Level 1 still live in societies where it can be difficult to intervene rapidly against an aggressively spreading disease. We need to ensure that basic health care reaches everyone, everywhere, so that outbreaks can be discovered more quickly. And we need the World Health Organization to remain healthy and strong to coordinate a global response.

Financial Collapse

In a globalized world, the consequences of financial bubbles are devastating. They can crash the economies of entire countries and put huge numbers of people out of work, creating disgruntled citizens looking for radical solutions. A really large bank collapse could be way worse than the global eruption that started with the US housing loan crash in 2008. It could crash the entire global economy.

Since even the best economists in the world failed to predict the last crash and fail year on year to predict the recovery from it—because the system is too complicated for accurate predictions—there is no reason to suppose that because no one is predicting a crash, it will not happen. If we had a simpler system there might be some chance of understanding it and working out how to avoid future collapses.

World War III

My whole life I have done all I can to establish relations with people in other countries and cultures. It's not only fun but also necessary to strengthen the global safety net against the terrible human instinct for violent retaliation and the worst evil of all: war.

We need Olympic Games, international trade, educational exchange programs, free internet—anything that lets us meet across ethnic groups and country borders. We must take care of and strengthen our safety nets for world peace. Without world peace, none of our sustainability goals will be achievable. It's a huge diplomatic challenge to prevent the proud and nostalgic nations with a violent track record from attacking others now that they are losing their grip on the world market. We must help the old West to find a new way to integrate itself peacefully into the new world.

Climate Change

It is not necessary to look only at the worst-case scenario to see that climate change poses an enormous threat. The planet's common resources, like the atmosphere, can only be governed by a globally respected authority, in a peaceful world abiding by global standards.

This can be done: we did it already with ozone depleters and with lead in gasoline, both of which the world community reduced to almost zero in two decades. It requires a strong, well-functioning international community (to be clear, I am talking about the UN). And it requires some sense of global solidarity toward the needs of different people on different income levels. The global community cannot claim such solidarity if it talks about denying the 1 billion people on Level 1 access to electricity, which would add almost nothing to overall emissions. The richest countries emit by far the most CO_2 and must start improving first before wasting time pressuring others.

Extreme Poverty

The other risks I have mentioned are highly probable scenarios that would bring unknown levels of future suffering. Extreme poverty isn't really a risk. The suffering it causes is not unknown, and not in the future. It's a reality. It's misery, day to day, right now. It is also where Ebola outbreaks come from, because there are no health services to encounter them at an early stage; and where civil wars start, because young men desperate for food and work, and with nothing to lose, tend to be more willing to join brutal guerrilla movements. It's a vicious circle: poverty leads to civil war, and civil war leads to poverty. The civil conflicts in Afghanistan and central Africa mean that all other sustainability projects in those places are on hold. Terrorists hide in the few remaining areas of extreme poverty. When rhinos are stuck in the middle of a civil war, it's much more difficult to save them.

Today, a period of relative world peace has enabled a growing global prosperity. A smaller proportion of people than ever before is stuck in extreme poverty. But there are still at least 800 million people left. Unlike with climate change, we don't need predictions and scenarios. We know that 800 million are suffering right now. We also know the solutions: peace, schooling, universal basic health care, electricity, clean water, toilets, contraceptives, and microcredits to get market forces started. There's no innovation needed to end poverty. It's all about walking the last mile with what's worked everywhere else. And we know that the quicker we act, the smaller the problem, because as long as people remain in extreme poverty they keep having large families and their numbers keep increasing. Providing these necessities of a decent life, quickly, to the final billion is a clear, fact-based priority.

The hardest to help will be those stuck behind violent and chaotic armed gangs in weakly governed states. To escape poverty, they will need a stabilizing military presence of some kind. They will need police officers with guns and government authority to defend innocent citizens against violence and to allow teachers to educate the next generation in peace.

Still I'm possibilistic. The next generation is like the last runner in a very long relay race. The race to end extreme poverty has been a marathon, with the starter gun fired in 1800. This next generation has the unique opportunity to complete the job: to pick up the baton, cross the line, and raise its hands in triumph. The project must be completed. And we should have a big party when we are done.

Knowing that some things are enormously important is, for me, relaxing. These five big risks are where we must direct our energy. These risks need to be approached with cool heads and robust, independent data. These risks require global collaboration and global resourcing. These risks should be approached through baby steps and constant evaluation, not through drastic actions. These risks should be respected by all activists, in all causes. These risks are too big for us to cry wolf.

I don't tell you not to worry. I tell you to worry about the right things. I don't tell you to look away from the news or to ignore the activists' calls to action. I tell you to ignore the noise, but keep an eye on the big global risks. I don't tell you not to be afraid. I tell you to stay coolheaded and support the global collaborations we need to reduce these risks. Control your urgency instinct. Control all your dramatic instincts. Be less stressed by the imaginary problems of an overdramatic world, and more alert to the real problems and how to solve them.

Factfulness

Factfulness is . . . recognizing when a decision feels urgent and remembering that it rarely is.

To control the urgency instinct, **take small steps.**

- **Take a breath.** When your urgency instinct is triggered, your other instincts kick in and your analysis shuts down. Ask for more time and more information. It's rarely now or never and it's rarely either/or.

- **Insist on the data.** If something is urgent and important, it should be measured. Beware of data that is relevant but inaccurate, or accurate but irrelevant. Only relevant and accurate data is useful.

- **Beware of fortune-tellers.** Any prediction about the future is uncertain. Be wary of predictions that fail to acknowledge that. Insist on a full range of scenarios, never just the best or worst case. Ask how often such predictions have been right before.

- **Be wary of drastic action.** Ask what the side effects will be. Ask how the idea has been tested. Step-by-step practical improvements, and evaluation of their impact, are less dramatic but usually more effective.

CHAPTER ELEVEN

FACTFULNESS IN PRACTICE

How Factfulness Saved My Life

"I think we should run," whispered the young teacher standing beside me.

Two thoughts raced across my mind. One was that if the teacher took off, I would have no way of communicating with the agitated crowd in front of me. I grabbed his arm and held on tightly.

The other thought was something that a wise governor of Tanzania had told me: "When someone threatens you with a machete, never turn your back. Stand still. Look him straight in the eye and ask him what the problem is."

It was 1989 and I was in a remote and extremely poor village named Makanga in the Bandundu region of what was then Zaire and is now the Democratic Republic of Congo. I was part of a team investigating an epidemic of the incurable paralytic disease called konzo that I had first discovered in Mozambique years earlier.

The research project had been two years in the planning and everything—all the approvals, drivers, translators, and lab equipment—had been meticulously prepared. But I had made one serious mistake. I had not explained properly to the villagers what

I wanted to do and why. I wanted to interview all the villagers and take samples of their food, and their blood and urine, and I should have been with the head of the village when he explained that to them.

That morning, as I had been quietly and methodically setting up in the hut, I heard villagers starting to gather outside. They somehow seemed uneasy but I was occupied with getting the blood sample machine to work. Eventually I managed to start the diesel generator and do a test run with the centrifuge. The machines were noisy and it was only when I switched them off that I heard the raised voices. Things had changed in seconds. I bent forward and stepped out of the low door. It had been dark in the hut and when I straightened up at first I couldn't see a thing. Then I saw: a crowd of maybe 50 people, all upset and angry. Some of them were pointing their fingers at me. Two men raised muscled arms and waved big machete knives.

That was when the teacher, my translator, suggested we run. I looked right and left and saw nowhere to go. If the villagers really wanted to hurt me there were enough of them to hold me back and let the machete men cut me down.

"What's the problem?" I asked the teacher.

"They are saying that you are selling the blood. You are cheating us. You are giving money only to the chief, and then you are going to make something with the blood that will hurt us. They say you shouldn't steal their blood."

This was very bad. I asked him if he would translate for me and then I turned to the crowd. "Can I explain?" I asked the villagers. "I can either leave your village right away, if you want, or I can explain why we have come."

"Tell us first," the people said. (Life is boring in these remote villages, so they probably thought, We can let him talk first, and we can kill him afterward.) The crowd held back the men with the machetes: "Let him talk."

This was the talk we should have had before. If you want to go into a village to do research, you have to take small steps, take your time, and be respectful. You have to let people ask all their questions, and you have to answer them.

I started to explain that we were working on a disease named konzo. I had photos from Mozambique and Tanzania, where I had studied konzo before, which I showed them. They were very interested in the photos. "We think it's linked to how you prepare the cassava," I said.

"No, no, no," they said.

"Well, we want to do this research, to test whether we are right. If we can find out, maybe you won't get the disease anymore."

Many of the children in the village had konzo. We had noticed them when we first arrived, lagging behind while the other children ran alongside our jeep with charming curiosity. I had spotted some children in this crowd with the classic spastic walking style too.

People began to mumble. One of the machete men, the more dangerous-looking one, with bloodshot eyes and a big scar down his forearm, started screaming again.

And then a barefoot woman, perhaps 50 years old, stepped out of the crowd. She strode toward me and then turned, threw out her arms, and in a loud voice said, "Can't you hear that it makes sense, what he is saying? Shut up! It makes sense. This blood test is necessary. Don't you remember everyone who died from measles? So many of our children died. Then they came and gave the children the vaccine, remember, and now no child ever dies of that disease. OK?"

The crowd shouted back, unmollified. "Yes, measles vaccine was good. But now they want to come take our blood."

The woman paused, then took a step toward the crowd. "How do you think they discovered the measles vaccine? Do you think it grows on trees in their countries? Do you think they pulled it out of the ground? No, they do what this doctor calls"—and she looked at me—"RE-SEAR-CHE." As she repeated the word the translator had used for research, she turned round and pointed at me. "That is how they find out how to cure diseases. Don't you see?"

We were in the most remote part of Bandundu, and here this woman had stepped up like the secretary of the Academy of Science and defended scientific research.

"I have a grandchild crippled for life by this *konzo*. The doctor says he can't cure it. But if we let him study us, perhaps he will find a way to stop it, like they stopped measles, so that we don't have to see our children and our grandchildren crippled anymore. This makes sense to me. We, the people of Makanga, need this 'research.'" Her dramatic talent was amazing. But she didn't use it to distort the facts. She used it to explain them. Forcefully, in a manner I had seen confident African women act in villages many times before, she rolled up her left sleeve. She turned her back on the crowd, pointed with her other hand to the crook of her arm, and looked me in the eyes. "Here. Doctor. Take my blood."

The men with machetes lowered their arms and moved away. Five or six others wandered off, grumbling. Everyone else lined up behind the woman to give their blood, the shouting replaced by soft voices and faces turned from anger to curious smiles.

I have always been extremely thankful for this courageous woman's insight. And now that we have defined Factfulness after years of fighting ignorance, I am amazed at how well it describes her behavior. She seemed to recognize all the dramatic instincts that had been triggered in that mob, helped them gain control over them, and convinced her fellow villagers with rational arguments. The fear instinct had been triggered by the sharp needles, the blood, and the disease. The generalization instinct had put me in a box as a plundering European. The blame instinct made the villagers take a stand against the evil doctor who had come to steal their blood. The urgency instinct made people make up their minds way too fast.

Still, under this pressure, she had stood up and spoken out. It had nothing to do with formal education. She most certainly had never left Bandundu and I'm sure she was illiterate. Without a doubt she had never learned statistics or spent time memorizing facts about the world. But she had courage. And she was able to think critically and express herself with razor-sharp logic and perfect rhetoric at a moment of extreme tension. Her factfulness saved my life. And if that woman could be factful under those circumstances, then you, highly educated, literate reader who just read this book, you can do it too.

Factfulness in Practice

How can you use Factfulness in your everyday life: in education, in business, in journalism, in your own organization or community, and as an individual citizen?

Education

In Sweden we don't have volcanoes, but we have geologists who are paid out of public funds to study volcanoes. Even regular schoolkids learn about volcanoes. Here in the Northern Hemisphere, astronomers learn about stars that can be seen only in the Southern Hemisphere. And at school, children learn about these stars. Why? Because they are part of the world.

Why then do our doctors and nurses not learn about the disease patterns on every income level? Why are we not teaching the basic up-to-date understanding of our changing world in our schools and in corporate education?

We should be teaching our children the basic up-to-date, fact-based framework—life on the four levels and in the four regions—and training them to use Factfulness rules of thumb—the bullet points from the end of each chapter. This would enable them to put the news from around the world in context and spot when the media, activists, or salespeople are triggering their dramatic instincts with overdramatic stories. These skills are part of the critical thinking that is already taught in many schools. They would protect the next generation from a lot of ignorance.

- We should be teaching our children that there are countries on all different levels of health and income and that most are in the middle.
- We should be teaching them about their own country's socioeconomic position in relation to the rest of the world, and how that is changing.
- We should be teaching them how their own country progressed through the income levels to get to where it is now, and how to use that knowledge to understand what life is like in other countries today.
- We should be teaching them that people are moving up the income levels and most things are improving for them.
- We should be teaching them what life was really like in the past so that they do not mistakenly think that no progress has been made.
- We should be teaching them how to hold the two ideas at the same time: that bad things are going on in the world, but that many things are getting better.
- We should be teaching them that cultural and religious stereotypes are useless for understanding the world.
- We should be teaching them how to consume the news and spot the drama without becoming stressed or hopeless.
- We should be teaching them the common ways that people will try to trick them with numbers.

- We should be teaching them that the world will keep changing and they will have to update their knowledge and worldview throughout their lives.

Most important of all, we should be teaching our children humility and curiosity.

Being humble, here, means being aware of how difficult your instincts can make it to get the facts right. It means being realistic about the extent of your knowledge. It means being happy to say "I don't know." It also means, when you do have an opinion, being prepared to change it when you discover new facts. It is quite relaxing being humble, because it means you can stop feeling pressured to have a view about everything, and stop feeling you must be ready to defend your views all the time.

Being curious means being open to new information and actively seeking it out. It means embracing facts that don't fit your worldview and trying to understand their implications. It means letting your mistakes trigger curiosity instead of embarrassment. "How on earth could I be so wrong about that fact? What can I learn from that mistake? Those people are not stupid, so why are they using that solution?" It is quite exciting being curious, because it means you are always discovering something interesting.

But the world will keep changing, and the problem of ignorant grown-ups will not be solved by teaching the next generation. What you learn about the world at school will become outdated within 10 or 20 years of graduating. So we must find ways to update adults' knowledge too. In the car industry, cars are recalled when a mistake is discovered. You get a letter from the manufacturer saying, "We would like to recall your vehicle and replace the brakes." When the facts about the world that you were taught in schools and universities become out of date, you should get a letter too: "Sorry, what we taught you is no longer true. Please return your brain for a free upgrade."

Or perhaps your employer should tackle the problem: "Please go through this material and take this test, to avoid embarrassing yourself at the World Economic Forum or similar."

Replace Sombreros with Dollar Street

Children start learning about other countries and religions in preschool. Cute little world maps with people in folklore dress from across the world are intended to make them aware of and respectful toward other cultures. The intention is good but these kinds of illustrations can create an illusion of great difference. People in other countries can seem stuck in historic and exotic ways of life. Of course some Mexicans sometimes wear large sombreros, but these large hats nowadays are probably more common on the heads of tourists.

Let's show children Dollar Street instead, and show them how regular people live. If you are a teacher, send your class "traveling" on dollarstreet.org and ask them to find differences within countries and similarities across countries.

Business

A single typo in your CV and you probably don't get the job. But if you put 1 billion people on the wrong continent you can still get hired. You can even get a promotion.

Most Western employees in large multinationals and financial institutions are still trying to operate according to a deeply rooted, outdated, and distorted worldview. Yet global understanding is becoming more and more crucial, and more and more possible. Most of us now work with consumers, producers, service providers, colleagues, or clients all across the planet. Some decades ago, when it was perhaps less important for us to know about the world, there were almost no reliable and accessible global statistics. As the world changed, though, the need for knowledge about the world also changed. Today, reliable data is easily available for almost every subject. This is quite new: my first partner in the fight against mega misconceptions was a photocopier, but today all that data is freely available online. In recruitment, production, marketing, and investment, it has never been easier or more important for business leaders and employees to act on a fact-based worldview.

Using data to understand the globalized markets has already become part of the culture. But when people's worldviews are upside down, data snippets can be just as misleading as wrong data or no data. Until one day someone actually tests their global knowledge, everyone assumes they are getting it kind of right.

In sales and marketing, if you run a big business in Europe or the United States, you and your employees need to understand that the world market of the future will be growing primarily in Asia and Africa, not at home.

In recruitment, you need to understand that being a European or US company no longer gives you bragging rights to attract international employees. Google and Microsoft, for example, have become global businesses and made their "Americanness" almost invisible. Their employees in Asia and Africa want to be part of truly global companies and they are. Their CEOs, Sundar Pichai of Google and Satya Nadella of Microsoft, were both raised and educated in India.

When I present to European corporations, I always tell them to tune down their European branding ("remove the Alps from your logo") and to move their headquarters—but not their European staff—elsewhere.

In production, you need to understand that globalization is not over. Decades ago, Western companies realized that industrial production had to be outsourced to the so-called emerging markets on Level 2, where products could be manufactured at the same quality for less than half the price. However, globalization is a continuing process, not a one-off event. The textiles industry that moved from Europe to Bangladesh and Cambodia as they reached Level 2 some decades ago will most likely soon move again as Bangladesh and Cambodia become wealthier and approach Level 3. These countries will have to diversify or suffer the consequences as their textiles jobs are shifted to African countries.

In making investment decisions, you need to shake off any naïve views of Africa shaped by the colonial past (and maintained by today's media) and understand that Ghana, Nigeria, and Kenya are where some of the best investment opportunities can be found today.

I think it will not be long before businesses care more about fact mistakes than they do about speling miskates, and will want to ensure their employees and clients are updating their worldview on a regular basis.

Journalists, Activists, Politicians

Journalists, activists, and politicians are also humans. They are not lying to us. They suffer from a dramatic worldview themselves. Like everyone else, they should regularly check and update their worldview and develop factful ways of thinking.

There are further actions that journalists can take to help them to present a less distorted worldview to the rest of us. Setting numbers in their historical context can help to keep them in proportion. Some journalists, aware of the distorting influence of negative news, are outlining new standards for more constructive news, with the goal of changing bad news habits and making journalism more meaningful. It's hard to tell at this point how much impact they will have.

Ultimately, it is not journalists' role, and it is not the goal of activists or politicians, to present the world as it really is. They will always have to compete to engage our attention with exciting stories and dramatic narratives. They will always focus on the unusual rather than the common, and on the new or temporary rather than slowly changing patterns.

I cannot see even the highest-quality news outlets conveying a neutral and nondramatic representative picture of the world, as statistics agencies do. It would be correct but just too boring. We should not expect the media to move very far in that direction. Instead it is up to us as consumers to learn how to consume the news more factfully, and to realize that the news is not very useful for understanding the world.

Your Organization

Once a year, the ministers of health from every country come together at the World Health Assembly. They plan health systems and compare health outcomes of different countries and then they have coffee. One time, the minister of health from Mexico whispered in my ear during a coffee break, "I care a lot about Mexico's average number, one day every year. That is today. All the other 364 days I only care about the differences within Mexico."

In this book, I have discussed ignorance of facts on a global level. I think there must be systematically ignored facts on the country level too, and in every community and every organization.

So far we have only tried a few local fact questions, but it seems like they follow a very similar pattern to the global facts we have tested more widely. In Sweden, for example, we asked:

Today, 20 percent of Swedes are older than 65. What will the number be 10 years from now:

- ☐ A: 20 percent
- ☐ B: 30 percent
- ☐ C: 40 percent

The correct answer is 20 percent—no change—but only 10 percent of Swedes picked that answer. That is devastating ignorance about a basic fact that is crucial in our Swedish debate about planning for the next ten years. I think it is because people have heard a lot about the aging population over the last 20 years, when the number did in fact increase, and then they assume a straight line.

There are so many more local and subject area fact questions we would love to try. Do people in your city know the basic proportions and trends that are shaping the future of the place they live in? We don't know, because we haven't tested it. But most likely: no.

What about your niche of expertise? If you work on marine life around Scandinavia, do your colleagues know the basic facts about the Baltic Sea? If you work in forestry, do your colleagues know if wildfires are getting more or less common? Do they know whether the latest fires caused more or less damage than those in the past?

We think there are endless such ignorances to discover if the fact questions are asked. Which is exactly why we suggest that as step one. You can hunt for ignorance in your own organization using the same methods we have used. Start simply by asking what are the most important facts in your organization and how many people know them.

Sometimes people get nervous about this. They think their colleagues and friends will be offended if they start checking their knowledge, and will not appreciate being proved wrong. My experience is the opposite. People like it a lot. Most people find it inspiring to realize what the world looks like. Most people are eager to start learning. Testing their knowledge, if it is done in a humble way, can release an avalanche of curiosity and new insights.

Final Words

I have found fighting ignorance and spreading a fact-based worldview to be a sometimes frustrating but ultimately inspiring and joyful way to spend my life. I have found it useful and meaningful to learn about the world as it really is. I have found it deeply rewarding to try to spread that knowledge to other people. And I have found it so exciting to finally start to understand why spreading that knowledge and changing people's worldviews have been so damn hard.

Could everyone have a fact-based worldview one day? Big change is always difficult to imagine. But it is definitely possible, and I think it will happen, for two simple reasons. First: a fact-based worldview is more useful for navigating life, just like an accurate GPS is more useful for finding your way in the city. Second, and probably more important: a fact-based worldview is more comfortable. It creates less stress and hopelessness than the dramatic worldview, simply because the dramatic one is so negative and terrifying.

When we have a fact-based worldview, we can see that the world is not as bad as it seems—and we can see what we have to do to keep making it better.

FACTFULNESS RULES OF THUMB

1. GAP

Look for the majority

2. NEGATIVITY

Expect bad news

3. STRAIGHT LINE

Lines might bend

4. FEAR

Calculate the risks

5. SIZE

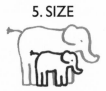

Get things in proportion

6. GENERALIZATION

Question your categories

7. DESTINY

Slow change is still change

8. SINGLE

Get a tool box

9. BLAME

Resist pointing your finger

10. URGENCY

Take small steps

OUTRO

In September 2015, Hans and the two of us decided to write a book together. On February 5, 2016, Hans received a diagnosis of incurable pancreatic cancer. The prognosis was bad. Hans was given two or three months to live or, if the palliative treatments were very successful, perhaps one year.

After the initial horrible shock, Hans thought things through. Life would continue for a while. He would still be able to enjoy time with his wife, Agneta, and his family and friends. But day-to-day, his health would be unpredictable. So within a week he had canceled all his 67 planned lectures for the coming year, as well as all planned TV and radio appearances and film productions. Hans was so sad to do it, but he realized he had no choice. And this dramatic change to his professional life was made bearable by one thing: the book. Following the diagnosis there was pleasure in the sadness as the book turned from being a burden on top of other tasks to being Hans's intellectual inspiration and joy.

There was so much he wanted to say. Over the next months, in our enthusiasm, the three of us pulled together enough material for a very thick book: about Hans's life, the work we had done together, and our latest ideas. Until the very end, he remained curious and passionate about the world.

We agreed on the outline for the book and started to write it. We had worked together on challenging projects for many years, and were used to constantly fighting over how best to explain a particular fact or concept. We were quickly humbled to discover how easy the collaboration had been during the years when we had all been well, and how terribly difficult it was to maintain our usual sharp and combative way of working now that Hans was ill. We almost failed.

On the evening of Thursday, February 2, 2017, Hans's health suddenly deteriorated. An ambulance was called, and into it Hans took printed copies of several chapters of the latest draft, his scribbled notes all over them. Four days later, in the early hours of Tuesday, February 7, Hans died. He had taken comfort over those last days from the drafts, discussing them with Ola from his hospital bed and dictating an email to the publishers, which said that he thought we had at last achieved "exactly the kind of book we have been aiming for." "Our joint work," Hans wrote, "is finally being turned into an enjoyable text that will help a global audience to understand the world."

When we announced Hans's death, an avalanche of condolences immediately poured in from friends, colleagues, and admirers from all over the world. Tributes to Hans were all over the internet. Our family and friends organized a ceremony at Karolinska Institutet and a funeral at Uppsala Castle, which together beautifully reflected the Hans we knew: brave, innovative, and serious-minded, yet always looking for the circus around the corner; a great friend and colleague and a beloved family member. The circus was there. There was a sword swallower onstage, of course (Hans's friend, whose X-ray you saw at the beginning of this book) and our son Ted did his own homemade trick with a bandy stick and helmet. (Bandy is a bit like ice hockey but friendlier.) We concluded with Frank Sinatra's anthem "My Way." Not just because Hans always did it His Way, but because of a lucky accident of a few years earlier. Hans didn't care much about music and he always insisted he was totally tone deaf, but his youngest son, Magnus, had once heard him sing. Hans had accidentally called Magnus from his pocket and, completely unaware, left him a four-minute voice message. This recorded Hans driving through traffic while singing loudly and lustily to Frank Sinatra's defiant anthem. This was just so Hans. You have seen his list of global risks but it couldn't stop him from singing on his way to work. Two thoughts at the same time: concerned and full of joy.

We had worked with Hans for 18 years. We had written his scripts and directed his TED talks, and argued with him for hours (sometimes months) about every detail of them. We had heard all his stories many times and had them recorded in many forms.

Working on the book had been painful in the last months of Hans's life but was strangely comforting in the months immediately after his death. As we completed this precious task, Hans's voice was always in our heads, and we often felt that he was not gone but still in the room beside us. Finishing the book felt like the best way to keep him with us and to honor his memory.

Hans would have loved promoting this book, and he would have done it brilliantly, but he knew from the moment of his diagnosis that that was not going to be possible. Instead, it falls to us to continue his mission and ours. Hans's dream of a fact-based worldview lives on in us and, we hope now, in you too.

Anna Rosling Rönnlund and Ola Rosling
Stockholm, 2018

ACKNOWLEDGMENTS

Most of what I understand about the world I learned not from studying data or sitting in front of a computer reading research papers—though I have done a lot of that too—but from spending time with, and discussing the world with, other people. I have had the privilege of traveling, studying, and working all over the world, with people from every continent, every major world religion, and, most importantly, at all income levels. I have learned a lot from the CEOs of international businesses and from my PhD students in Stockholm. I have learned even more from women living in extreme poverty in Africa; from Catholic nuns working in the most remote villages; from medical students in Bangalore and academics from Nigeria, Tanzania, Vietnam, Iran, and Pakistan; and from the thought leaders of countries on all income levels, from Eduardo Mondlane to Melinda Gates. I want to thank all of you for sharing your knowledge with me, for making my life so rich and wonderful, and for showing me a world completely different from the one I learned about in school.

Understanding the world is one thing. Turning that understanding into a book is another. As always, it is the team behind the scenes who make it possible. Thank you to each one of the dedicated and creative members of staff at Gapminder who built the resources that I used in all my lectures.

Thank you to our literary agent, Max Brockman, for great advice and support, and to our editors, Drummond Moir at Hodder in the United Kingdom and Will Schwalbe at Macmillan in the United States, for believing in the book, for their kind and calm guidance through the process, and for their wise counsel on how to improve the book. Thanks too to Harald Hultqvist for telling us we had to get an international agent, and to Richard Herold, our editor in Sweden, for being an excellent adviser through the early process and throughout. Thanks to Bill Warhop, the copy editor, and Bryn Clark, for their input. If you found this book readable, it is thanks to Deborah Crewe. She was brave to take on three authors with way too much material. She listened hard to what we wanted, and then worked patiently and with great skill, speed, and humor, turning our eccentric Swenglish into what you have just read. Even more important, she was able to absorb our thousands of fact snippets, anecdotes, and rules of thumb, and help us to mold them into one coherent epic. We are so grateful to our new dear friend.

Special thank-yous to Max, Ted, and Ebba for letting me spend so many weekends and evenings with your parents, Anna and Ola. I hope that when you read this book and see the work we have been doing you will forgive me a little. And thank you for your own contributions: to Max (12), who spent hours discussing research with me in my office and editing hundreds of my recorded transcripts; to Ted (10), who took photos for Dollar Street, tested our fact questions on his classmates, and went to New York to receive the UN Population Award on my behalf; and to Ebba (8), who came up with clever ideas on how to improve the material and design the artwork you see throughout the book.

There is a phrase in Swedish, "stå ut." It means putting up with, bearing with, enduring, hanging in there. I hope my family, friends, and colleagues know how grateful I am that they have "stått ut" with me so much over the years. I realize that the way I work—the way I am—means I have often been absent or, if not absent, then rushing in and out. I know that even when I have been present I have often been distracted and irritating. I can be a frustrating person when I am working, which is almost all the time I am awake. So my thanks go to everyone I have the honor to call a friend and colleague. It is hard to pick out one friend and colleague above all others but I must particularly thank Hans Wigzell, who bravely supported Gapminder from the very beginning and who was with me until the last day, tirelessly trying to figure out ways to prolong my life.

Above all, for their endless patience and love, my deep and sincere thanks go to my teenage love, wife, and companion throughout my life, Agneta; to my beloved children, Anna, Ola, and Magnus, and their spouses; and to my grandchildren, Doris, Stig, Lars, Max, Ted, Ebba, Tiki, and Mino, who every day give me hope for the future.

Ola, Anna, and I would also like to thank:

Jörgen Abrahamsson, Christian Ahlstedt, Johan Aldor, Chris Anderson, Ola Awad, Julia Bachler, Carl-Johan Backman, Shaida Badiee, Moses Badio, Tim Baker, Ulrika Baker, Jean-Pierre Banea-Mayambu, Archie Baron, Aluisio Barros, Luke Bawo, Linus Bengtsson, Omar Benjelloun, Lasse Berg, Anna Bergström, Staffan Bergström, Anita Bergsveen, BGC3, the Bill and Melinda Gates Foundation, Sali Bitar, Pelle Bjerke, Stefan Blom, Anders Bolling, Staffan Bremmer, Robin Brittain-Long, Peter Byass, Arthur Câmara, Peter Carlsson, Paul Cheung, Sung-Kyu Choi, Mario Cosby, Andrea Curtis, Jörn Delvert, Kicki Delvert, Alisa Derevo, Nkosazana Dlamini-Zuma, Mohammed Dunbar, Nelson Dunbar, Daniel Ek, Anna Mia Ekström, Ziad El-Khatib, Mats Elzén, Martin Eriksson, Erling Persson Foundation,

Peter Ewers, Mosoka Fallah, Ben Fausone, Per Fernström, Guenther Fink, Steven Fisher, Luc Forsyth, Anders Frankenberg, Haishan Fu, Minou Fuglesang, Bill Gates, Melinda Gates, George Gavrilis, Anna Gedda, Ricky Gevert, Marcus Gianesco, Nils Petter Gleditsch, Google, Google Public Data team, Georg Götmark, Erik Green, Ann-Charlotte Gyllenram, Catharina Hagströmer, Sven Hagströmer, Nina Halden, Rasmus Hallberg, Esther Hamblion, Mona Hammami and the team in Abu Dhabi behind Looking Ahead, Katie Hampson, Hans Hansson, Per Heggenes, David Herdies, Dan Hillman, Mattias Högberg, Ulf Högberg, Magnus Höglund, Adam Holm, Anu Horsman, Matthias Horx, Abbe Ibrahim, IHCAR, IKEA foundation, Dikena G. Jackson, Oskar Jalkevik and his team at Transkribering.nu, Kent Janer, Jochnick Foundation, Claes Johansson, Jan-Olov Johansson, Klara Johansson, Jan Jörnmark, Åsa Karlsson, Linley Chiwona Karltun, Alan Kay, Haris Shah Khattak, Tariq Khokhar, Niclas Kjellström-Matseke, Tom Kronhöffer, Asli Kulane, Hugo Lagercrantz, Margaret Orunya Lamunu, Staffan Landin, Daniel Lapidus, Anna Rosling Larsson, Jesper Larsson, Pali Lehohla, Martin Lidholt, Victor Lidholt, Henrik Lindahl, Mattias Lindblad, Mattias Lindgren, Lars Lindkvist, Ann Lindstrand, Per Liss, Terence Lo, Håkan Lobell, Per Löfberg, Anna Mariann Lundberg, Karin Brunn Lundgren, Max Lundkvist, Rafael Luzano, Marcus Maeurer, Ewa Magnusson, Lars Magnusson, Jacob Malmros, Niherewa Maselina, Marissa Mayer, Branko Milanović, Zoriah Miller, Katayoon Moazzami, Sibone Mocumbi, Anders Mohlin, Janet Rae Johnson Mondlane, Louis Monier, Abela Mpobela, Paul Muret, Chris Murray, Hisham Najam, Sahar Nejat, Martha Nicholson, Anders Nordström, Lennart Nordström, Marie Nordström, Tolbert Nyenswah, Johan Nystrand, Martin Öhman, Max Orward, Gudrun Østby, Will Page, Francois Pelletier, Karl-Johan Persson, Stefan Persson, Måns Peterson, Stefan Swartling Peterson, Thiago

Porto, Postcode Foundation, Arash Pournouri, Amir Rahnama, Joachim Retzlaff, Hannah Ritchie, Ingegerd Rooth, Anders Rönnlund, David Rönnlund, Quiyan Rönnlund, Thomas Rönnlund, Max Roser and The World in Data team, Magnus Rosling, Pia Rosling, Siri Aas Rustad, Love Sahlin, Xavier Sala-i-Martín, Fia-Stina Sandlund, Ian Saunders, Dmitriy Shekhovtsov and his Valor Software, Harpal Shergill, Sida, Jeroen Smits, Cosimo Spada, Katie Stanton, Bo Stenson, Karin Strand, Eric Swanson, Amirhossein Takian, Lorine Zineb Nora "Loreen" Talhaoui, Manuel Tamez, Andreas Forø Tollefsen, Edward Tufte, Thorkild Tylleskär, UNDP, Henrik Urdal, Bas van Leeuwen, the family of Johan Vesterlund, Cesar Victoria, Johan von Schreeb, Alem Walji, Jacob Wallenberg, Eva Wallstam, Rolf Widgren, John Willmoth, Agnes Wold, Fredrik Wollsén and his team, World Health Organization, World We Want Foundation, Danzhen You, Guohua Zheng, and Zhang Zhongxing.

Mattias Lindgren for compiling most of the Gapminder historic time series for the economy and demography. All my students and doctoral students, from whom I learned so much, all the teachers and students who welcomed us to their schools to help us test our materials, all the amazing consultants around the world who have helped us, Jimmy Wales and the voluntary editors on Wikipedia, and all the Dollar Street families and photographers.

The previous and current board members of the Gapminder Foundation for their wise and stable support: Hans Wigzell, Christer Gunnarsson, Bo Sundgren, Gun-Britt Andersson, and Helena Nordenstedt (who also helped with fact-checking). And our amazing Gapminder team, Mikael Arevius, Klara Elzvik, Olof Gränström, Jasper Heeffer, Gabriela Sá, and Angie Skazka, headed by Fernanda Drumond, who tirelessly continued to develop Gapminder's free teaching materials while we finished this book. And who also gave invaluable input to the manuscript!

And finally, our wonderful friends and families, for being patient with us and for helping us throughout the process. You know who you are. Without you this book would not have been possible. Thank you.

APPENDIX
How Did Your Country Do?

In 2017, the Gapminder Test launched. It consists of 13 questions, all with an A, B, C alternative. In 2017, Gapminder worked with Ipsos MORI and Novus to test 12,000 people in 14 countries. Their polls were conducted with online panels weighted to be representative of the adult populations. The test was conducted in Australia, Belgium, Canada, Finland, France, Germany, Hungary, Japan, Norway, South Korea, Spain, Sweden, the United Kingdom, and the United States. The 13 fact questions are freely available in multiple languages at www.gapminder.org/test/2017. Read more about the results here: www.gapminder.org/test/2017/results.

To learn more about the methodology of these polls and the supporting data behind the correct answers, see "Notes" on pages 276–279.

Education of girls in low-income countries

FACT QUESTION 1 RESULTS: percentage who answered correctly.
In all low-income countries across the world today, how many girls finish primary school?
(Correct answer: 60%.)

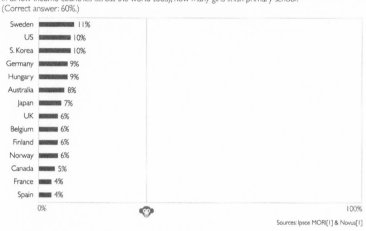

Sources: Ipsos MORI[1] & Novus[1]

Majority income level

FACT QUESTION 2 RESULTS: percentage who answered correctly.
Where does the majority of the world population live?
(Correct answer: middle-income countries.)

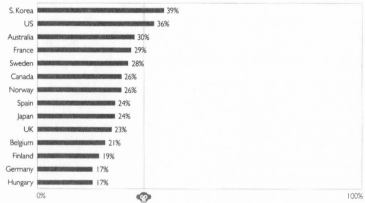

Sources: Ipsos MORI[1] & Novus[1]

Extreme poverty

FACT QUESTION 3 RESULTS: percentage who answered correctly.
In the last 20 years, the proportion of the world population living in extreme poverty has … ?
(Correct answer: almost halved.)

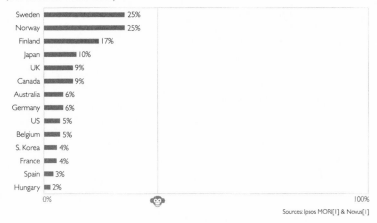

Sources: Ipsos MORI[1] & Novus[1]

Lifespan

FACT QUESTION 4 RESULTS: percentage who answered correctly.
What is the life expectancy of the world today?
(Correct answer: 70 years.)

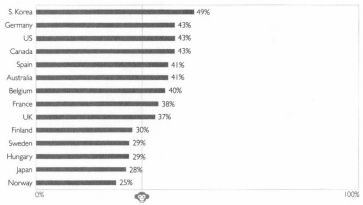

Sources: Ipsos MORI[1] & Novus[1]

Future number of children*

FACT QUESTION 5 RESULTS: percentage who answered correctly.
There are two billion children in the world today, aged 0 to 15 years old. How many children will there be in the year 2100 according to the United Nations? (Correct answer: two billion children.)

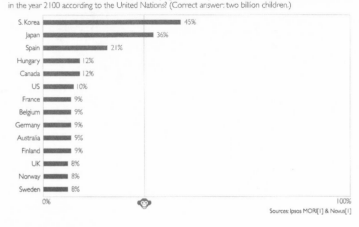

Sources: Ipsos MORI[1] & Novus[1]

More people

FACT QUESTION 6 RESULTS: percentage who answered correctly.
The UN predicts that by 2100 the world population will have increased by another 4 billion people. What is the main reason? (Correct answer: more adults.)

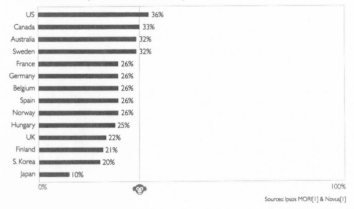

Sources: Ipsos MORI[1] & Novus[1]

---

* South Korea and Japan actually beat the chimps on this question. We don't know why yet. It could have to do with the skewed age structures in these countries. It could be that the fall in the birth rate is discussed more there than elsewhere. We have some more work to do to understand this.

Natural disasters

FACT QUESTION 7 RESULTS: percentage who answered correctly.

How did the number of deaths per year from natural disasters change over the last hundred years?
(Correct answer: decreased to less than half.)

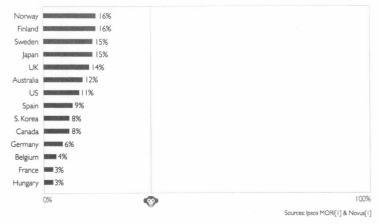

Country	Percentage
Norway	16%
Finland	16%
Sweden	15%
Japan	15%
UK	14%
Australia	12%
US	11%
Spain	9%
S. Korea	8%
Canada	8%
Germany	6%
Belgium	4%
France	3%
Hungary	3%

Sources: Ipsos MORI[1] & Novus[1]

Where people live

FACT QUESTION 8 RESULTS: percentage who answered correctly.

There are roughly 7 billion people in the world today. Which map shows best where they live?
(Correct answer: see the map.)

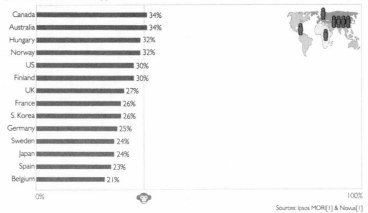

Country	Percentage
Canada	34%
Australia	34%
Hungary	32%
Norway	32%
US	30%
Finland	30%
UK	27%
France	26%
S. Korea	26%
Germany	25%
Sweden	24%
Japan	24%
Spain	23%
Belgium	21%

Sources: Ipsos MORI[1] & Novus[1]

Vaccination of children

FACT QUESTION 9 RESULTS: percentage who answered correctly.

How many of the world's 1-year-old children today have been vaccinated against some disease?
(Correct answer: 80%.)

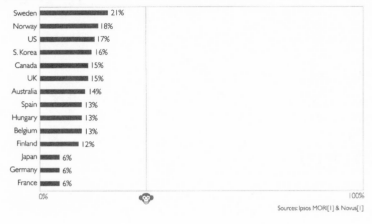

Sources: Ipsos MORI[1] & Novus[1]

Women's education

FACT QUESTION 10 RESULTS: percentage who answered correctly.

Worldwide, 30-year-old men have spent 10 years in school, on average. How many years have women
of the same age spent in school? (Correct answer: 9 years.)

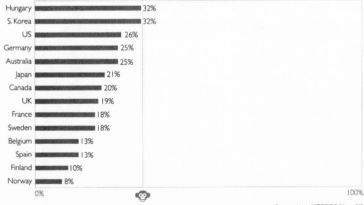

Sources: Ipsos MORI[1] & Novus[1]

Endangered animals

FACT QUESTION 11 RESULTS: percentage who answered correctly.

Tigers, giant pandas, and black rhinos were listed as threatened species in 1996. Since then, have any of these become more critically endangered? (Correct answer: none of them.)

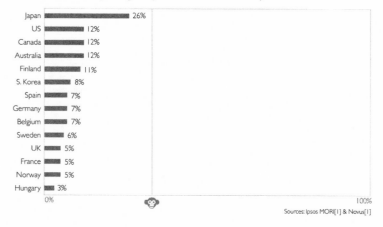

Sources: Ipsos MORI[1] & Novus[1]

Electricity

FACT QUESTION 12 RESULTS: percentage who answered correctly.

How many people in the world have some access to electricity? (Correct answer: 80%.)

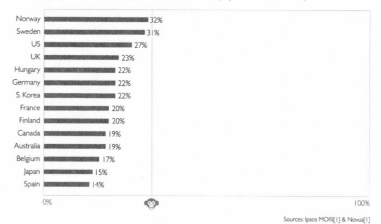

Sources: Ipsos MORI[1] & Novus[1]

Climate

FACT QUESTION 13 RESULTS: percentage who answered correctly.

Global climate experts believe that, over the next 100 years, the average temperature will...?
(Correct answer: ...get warmer.)

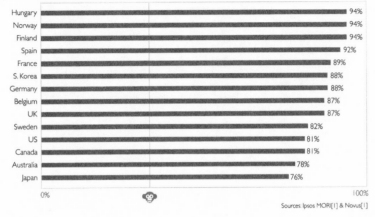

Sources: Ipsos MORI[1] & Novus[1]

Number of correct answers out of the first twelve questions

ONLY 10% ANSWERED BETTER THAN THE CHIMPS

Share of people by number of correct answers on the 12 questions.
(12,000 people in 14 countries.)

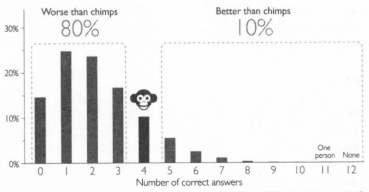

Sources: Novus[1], Ipsos MORI[1] and more info at gapm.io/rtest17

NOTES

We have taken enormous care to check and double-check our sources and the ways we have used them: in a book about Factfulness, we do not want to make a single fact mistake. But we are human beings; however hard we strive, we still make mistakes.

If you spot a mistake, please share your knowledge and enable us to improve this book. Contact us at factfulness-book@gapminder.org. And find all the mistakes that have already been spotted at: gapminder.org/factfulness /book/mistakes.

Below is a selected set of notes and sources. You can find the full list here: gapm.io/ffbn.

General Notes

Data for 2017. Throughout the book, where economic indicators do not extend to 2017, Gapminder has extended the series, mainly using forecasts from the World Economic Outlook from IMF[1]. For extending demographic data, we have used the World Population Prospect 2017, see UN-Pop[1]. See gapm.io/eext.

Country boundaries. Throughout the book, we refer to countries in the past as if they always had the boundaries they have today. For example, we talk about Bangladesh's family sizes and life expectancy in 1942 as if it had been an independent country at that time, although in reality it was still under British rule as part of British India. See gapm.io/geob.

Inside Cover

World Health Chart 2017. When you open the book, you see a colorful chart: the World Health Chart 2017. Each bubble is a country. The size of the bubble represents the country's population, and the color of the bubble its geographical region. On the x-axis is GDP per capita (PPP in constant 2011 international $) and on the y-axis life expectancy. The population data comes from UN-Pop[1], the GDP data from World Bank[1], and the life expectancy data from IHME[1], all extended to 2017 by Gapminder as described above. This chart, together with more information about the sources, is freely available at www.gapminder .org/whc.

Introduction

X-ray. The X-ray was taken by Staffan Bremmer at Sophiahemmet in Stockholm. The sword swallower is a friend of Hans's, called Maryanne Magdalen. Her website is here: gapm.io/xsword.
Fact questions. The 13 fact questions are freely available in multiple languages at www.gapminder.org/test/2017.
Online polls. Gapminder worked with Ipsos MORI and Novus to test 12,000 people in 14 countries. Their polls were conducted with online panels weighted to be representative of the adult populations—Ipsos MORI[1] and Novus[1]. The average number of correct answers for the 12 questions (i.e., excluding question 13 on climate change) was 2.2, which we rounded to 2. See more at gapm.io/rtest17.
Poll results. The results of the online polls by question and country are set out in the appendix. For the results of the polls we have conducted in our lectures, see gapm.io/rrs.
World Economic Forum lecture. For a video recording of the lecture (the audience receives its results five minutes and 18 seconds in), see WEF.
Fact Question 1: Correct answer is C. Sixty percent of the girls in low-income countries finish primary school. According to World Bank[3], the number is 63.2 percent, but we rounded it to 60 percent to avoid overstating progress. See gapm.io/q1.

Fact Question 2: Correct answer is B. The majority of people live in middle-income countries. The World Bank[2] divides countries into income groups based on gross national income per capita in current US $. According to the World Bank[4], the low-income countries represent 9 percent of the world population, the middle-income countries, 76 percent of the world population, and the high-income countries, 16 percent of the world population. See gapm.io/q2.

Fact Question 3: Correct answer is C. The share of people living on less than $1.9/day fell from 34 percent in 1993 to 10.7 percent in 2013, according to World Bank[5]. Despite the impression of precision given by the precise threshold of $1.9/day and the use of decimals, the uncertainties in these numbers are very large. Extreme poverty is very difficult to measure: the poorest people are mostly subsistence farmers or destitute slum dwellers, with unpredictable and constantly changing living conditions and few documented monetary transactions. But even if the exact levels are uncertain, the trend direction is not uncertain, because the sources of error are probably constant over time. We can trust that the level has fallen to at least half, if not one-third. See gapm.io/q3.

Fact Question 4: Correct answer is C. The average global life expectancy for those born in 2016 was 72.48 years, according to IHME[1]. The UN-Pop[3] estimate is slightly lower, at 71.9 years. We rounded to 70 to avoid overstating progress. See gapm.io/q4.

Fact Question 5: Correct answer is C. For the past ten years, UN-Pop[2] has published forecasts predicting that the number of children in the year 2100 will not be higher than it is today. See gapm.io/q5.

Fact Question 6: Correct answer is B. In their forecasts, the experts at the UN Population Division calculate that 1 percent of the population increase will come from 0.37 billion more children (age 0–14), 69 percent from 2.5 billion more adults (age 15 to 74), and 30 percent from 1.1 billion more very old people (age 75 and older). Data is from UN-Pop[3]. See gapm.io/q6.

Fact Question 7: Correct answer is C. Annual deaths from natural disasters have decreased by 75 percent over the past 100 years, according to the International Disaster Database; see EM-DAT. Since disasters vary from year to year, we compare ten-year averages. In the last ten years (2007–2016), on average 80,386 people were killed by natural disasters per year. This is 25 percent of the number 100 years earlier (1907–1916), when it was 325,742 deaths per year. See gapm.io/q7.

Fact Question 8: Correct answer is A. The world population in 2017 is 7.55 billion, according to UN-Pop[1]. That would usually be rounded to eight billion, but we show seven billion because we are rounding the population region by region. The populations of the four Gapminder[1] regions were estimated based on national data from UN-Pop[1]: the Americas, 1.0 billion; Europe, 0.84 billion; Africa, 1.3 billion; Asia, 4.4 billion. See gapm. io/q8.

Fact Question 9: Correct answer is C. Eighty-eight percent of one-year-old children in the world today are vaccinated against some disease, according to WHO[1]. We rounded it down to 80 percent to avoid overstating progress. See gapm.io/q9.

Fact Question 10: Correct answer is A. Worldwide, women aged 25 to 34 have an average of 9.09 years of schooling, and men have 10.21, according to IHME[2] estimates from 188 countries. Women aged 25 to 29 have an average of 8.79 years of schooling, and men 9.32 years, according to Barro and Lee (2013) estimates from 146 countries in 2010. See gapm.io /q10.

Fact Question 11: Correct answer is C. None of the three species are classified as more critically endangered today than they were in 1996, according to the IUCN Red List of Threatened Species. The tiger (*Panthera tigris*) was classified as Endangered (EN) in 1996, and it still is; see IUCN Red List[1]. But after a century of decline, tiger numbers in the wild are on the rise, according to WWF and Platt (2016). According to IUCN Red List[2], the giant panda (*Ailuropoda melanoleuca*) was classified as Endangered (EN) in 1996, but in 2015, new assessments of increasing wild populations resulted in a change of classification to the less critical status Vulnerable (VU). The black rhino (*Diceros bicornis*) was classified as Critically Endangered (CR) and still is; see IUCN Red List[3]. But the International Rhino Foundation says many populations in the wild are slowly increasing. See gapm.io/q11.

Fact Question 12: Correct answer is C. A majority of the world population, 85.3 percent, had some access to the electricity grid in their countries, according to GTF. We rounded this down to 80 percent to avoid overstating progress. The term "access" is defined differently in all their underlying sources. In some extreme cases, households may experience an average of 60 power outages per week and still be listed as "having access to electricity." The question, accordingly, talks about "some" access. See gapm.io/q12.

Fact Question 13: Correct answer is A. "Climate experts" refers to the 274 authors of the IPCC[1] Fifth Assessment Report (AR5), published in 2014 by the Intergovernmental Panel on Climate Change (IPCC), who write, "Surface temperature is projected to rise over the 21st century under all assessed emission scenarios"; see IPCC[2]. See gapm.io/q13.

Illusions. The idea of explaining cognitive biases using the Müller-Lyer illusion comes from *Thinking, Fast and Slow,* by Daniel Kahneman (2011).

The ten instincts and cognitive psychology. Our thinking on the ten instincts was influenced by the work of a number of brilliant cognitive scientists. Some of the books that completely changed our thinking about the mind and about how we should teach facts about the world are: Dan Ariely, *Predictably Irrational* (2008), *The Upside of Irrationality* (2010), and *The Honest Truth About Dishonesty* (2012); Steven Pinker, *How the Mind Works* (1997), *The Stuff of Thought* (2007), *The Blank Slate* (2002), and *The Better Angels of Our Nature* (2011); Carol Tavris and Elliot Aronson, *Mistakes Were Made (But Not by Me)* (2007); Daniel Kahneman, *Thinking, Fast and Slow* (2011); Walter Mischel, *The Marshmallow Test* (2014); Philip E. Tetlock and Dan Gardner, *Superforecasting* (2015); Jonathan Gottschall, *The Storytelling Animal* (2012); Jonathan Haidt, *The Happiness Hypothesis* (2006) and *The Righteous Mind* (2012); and Thomas Gilovich, *How We Know What Isn't So* (1991).

Chapter One: The Gap Instinct

Child mortality. The child mortality data used in the 1995 lecture came from UNICEF[1]. In this book we have updated the examples and use the new mortality data from UN-IGME.

Bubble charts. The bubble charts on family size and child survival rates in 1965 and 2017 use data from UN-Pop[1,3,4] and UN-IGME. An interactive version of the chart is freely available here: gapm.io/voutdwv.

Low-income countries. Gapminder has asked the public in the United States and Sweden how they imagine life in "low-income countries" or "developing countries." They systematically guessed numbers that would have been correct 30 or 40 years ago. See gapm.io/rdev.

The primary school completion rate for girls is below 35 percent in just three countries. But for all three, the uncertainty is high and the numbers are outdated: Afghanistan (1993), 15 percent; South Sudan (2011),

18 percent; Chad (2011), 30 percent. Three other countries (Somalia, Syria, and Libya) have no official number. The girls in these six countries suffer under severe gender inequality, but in total they make up only 2 percent of all girls of primary school age in the world, based on UN-Pop[4]. Note that in these countries, many boys are also missing school. See gapm.io/twmedu.

Income levels. The numbers of people on the four income levels have been defined by Gapminder[8] based on data from PovcalNet and forecasts from IMF[1]. Incomes are adjusted for Purchasing Power Parity $ 2011 from ICP. See gapm.io/fwlevels.

The graphs showing people distributed by income, comparing incomes in Mexico and the United States in 2016, are based on the same data, slightly adjusted to align with the shape of the distributions from the latest available national income surveys. Brazil's numbers come from World Bank[16], PovcalNet, slightly adjusted to better align with CETAD. See gapm.io/ffinex.

Throughout the book, when talking about personal income levels and countries' average incomes, we use a doubling scale. Doubling (or logarithmic) scales are used in many situations when comparing numbers across a large range, or when small differences between small numbers are as important as big differences between big numbers. It's a useful scale when it is not the size of the pay rise that matters, but the size of the rise in relation to what you had before. See gapm.io/esca.

"Developing countries." Here is the World Bank announcing its plan to phase out the use of the term "developing world," five months after I explicitly challenged its outdated terminology: https://blogs.worldbank .org/opendata/should-we-continue-use-term-developing-world. See World Bank[15].

Large parts of the UN still use the term "developing countries", but there's no common definition. The UN Statistics Division (2017) uses it for something it calls "statistical convenience". and finds it convenient to classify as many as 144 countries as developing (including Qatar and Singapore, two of the healthiest and richest countries on the planet).

Math scores. Part of the example is borrowed from Denise Cummins (2014).

Extreme poverty. The term "extreme poverty" has a set technical meaning: it means you have a daily income of less than $1.9/day. The term "poverty" in many countries on Level 4 is a relative term, and the "poverty line" may refer to the threshold for eligibility for social welfare or the official statistical measure of poverty in that country. In Scandinavia, the official poverty lines are 20 times higher than the poverty lines in the poorest countries, like Malawi, even after adjusting for the large differences in purchasing power; see World Bank[17]. The latest US census estimates that 13 percent of the population lives below its poverty line, putting it at approximately $20/day. The social and economic challenges of being among the poorest in a rich country should not be neglected (see World Bank[5]), but it is not the same thing as being extremely poor. See gapm.io/tepov.

Chapter Two: The Negativity Instinct

The environment. The statements about overfishing and the deterioration of the seas are based on UNEP[1], FAO[2] and Paul Collier, *The Plundered Planet* (2010), p. 160, and data for endangered species comes from IUCN Red List[4]. See gapm.io/tnplu.

Bar chart: Better, worse, or about the same? The bar chart mixes results from YouGov[1] and Ipsos MORI[1], as an identical question was asked in different countries. See gapm.io/rbetter.

When to trust the data. In this chapter we introduce the idea that you should never trust the data 100 percent. For Gapminder's guidelines on reasonable doubt for different kinds of data, see gapm.io/doubt.

Graph: Extreme poverty trend. Historians have tried to estimate the extreme poverty rate in 1820 using different methods, and their results differ widely. Gapminder[9] roughly estimates that 85 percent of the world population lived on Level 1 in 1800. The post-1980 data comes from PovcalNet. Gapminder[9] has extended the trend to 2017 extending PovcalNet with IMF[1] forecasts. The numbers in the text on the reductions in extreme poverty in China, India, Latin America, and elsewhere come from World Bank[5]. See gapm.io/vepovt.

Life expectancy. Life expectancy data is from IHME[1]. In 2016, only the Central African Republic and Lesotho had a life expectancy as low as 50 years. However, uncertainties are huge, especially on Levels 1 and 2. Learn how much data doubt you should have at gapm.io/blexd.

Deaths from starvation in Ethiopia. This number is an average of two sources, FRD and EM-DAT.

Lesotho. The citizens of Lesotho are usually referred to as the Basotho. Many Basotho also live outside Lesotho, but here we refer to those actually living in Lesotho.

Literacy. Historic literacy numbers for Sweden are from van Zanden[2] and OurWorldInData[2]. The literacy rate for India is from India Census 2011. Both in India today and in Sweden 100 years ago, "literacy" may only mean basic recognition of letters and the ability to parse text slowly. The figures do not imply an ability to understand advanced written messages. See gapm.io/tlit.

Vaccination. Vaccination data comes from WHO[1]. Even in Afghanistan, more than 60 percent of the one-year-olds today have received multiple vaccinations. None of these vaccines existed when Sweden was on Level 1 or 2, which is part of the reason lives were shorter in Sweden back then. See gapm.io/tvac.

32 improvements. The data behind each of the 32 line charts on pages 60–63, together with detailed documentation of how the many sources were used, can be found here: gapm.io/ffimp.

Guitars per capita. For more information about this chart, see gapm.io/tcminsg.

Historic child murders. In violent communities, children are not spared. Members of hunter-gatherer groups generally experienced lots of violence, as described in Gurven and Kaplan (2007), Diamond (2012), Pinker (2011), and OurWorldInData[5]. This doesn't mean all tribes of hunter-gatherers are the same. In situations of extreme poverty all across the world, many cultures have accepted the practice of infanticide, the killing of one's own children to reduce the number of mouths to feed in difficult times. This terrifying way of losing a child is just as painful as other ways, as consistently documented in traditional societies by anthropologists interviewing parents who had to kill a newborn; see Pinker (2011), pp. 417.

Educating girls. The data on girls' and boys' education comes from UNESCO[5]. Schultz (2002) describes clearly and in more detail how educating girls has proven to be one of the world's best-ever ideas.

Drownings. The data on drownings today comes from IHME[4,5]. Up until 1900, more than 20 percent of the victims of drownings were children below the age of ten. The Swedish Life Saving Society started lobbying for obligatory swimming practice in all schools, which together with other preventive actions reduced the number; see Sundin et al. (2005).

Catching up. Use the animated version of the World Health Chart to see how almost all countries are now catching up with Sweden (or select another country to compare), at www.gapminder.org/whc.

Chapter Three: The Straight Line Instinct

Ebola. The data on Ebola is from WHO[3]. The material Gapminder produced to try to communicate the urgency of the situation is at gapm.io /vebol.

Population forecasts. Population forecasts are based on UN-Pop[1,2,5]. The demography experts at the UN Population Division have been very accurate in their forecasts for many decades, even before modern computer modeling was possible. Their forecasts of the future number of children have stayed the same in the past four editions of the publication. Two billion children is a rounded number. The precise UN numbers are 1.95 billion for 2017 and 1.97 billion for 2100. For more on the quality of UN forecasts, see Nico Keilman (2010) and Bongaarts and Bulatao (2000). See gapm.io/epopf.

Historic population data. The line showing the world population from 8000 BC to today uses data from hundreds of different sources, compiled by the economic historian Mattias Lindgren. The sources listed under the chart are only the main sources. See gapm.io/spop.

Babies per woman. We use the term "babies per woman" for the statistical indicator "total fertility rate." We use UN-Pop[3] for post-1950 data and Gapminder[7], based on Mattias Lindgren's work, for the years before 1950. The dashed line after 2017 shows the UN medium fertility projection, expected to reach 1.96 in 2099. See gapm.io/tbab.

The fill-up. If you find it hard to understand the fill-up in the text and static images in this book, we find it easier to explain with animations, or with our own hands; see gapm.io/vidfu. (This phenomenon is also called the demographic momentum. For technical descriptions see UN-Pop[6, 7]). See gapm.io/efill.

Historic babies per woman and child mortality. The main sources behind our assumptions about fertility and mortality in pre-1800 families are Livi-Bacci (1989), Paine and Boldsen (2002), and Gurven and Kaplan (2007). Nobody knows the fertility rate before 1800, but six is a commonly used and likely average. See gapm.io/eonb.

Chart: Average family size by income. Our estimates for families on different income levels are based on household data compiled by Countdown to 2030 and GDL[1,2], combining hundreds of households surveys from UNICEF-MICS, USAID-DHS[1], IPUMS, and others. See Gapminder[30].

Changing the typical family size. For more on how societies transition from large to small families, see Rosling et al. (1992), Oppenheim Mason (1997), Bryant (2007), and Caldwell (2008). Babies per woman seems to start to increase again when people reach really high incomes on Level 4; see Myrskylä et al. (2009). This video shows how saving lives leads to fewer people: gapm.io/esclfp.

Straight lines, S-bends, slides, and humps. Most of these charts use national income data; see Gapminder[3]. A few (the straight line on recreational spending, the S-bend on vaccinations and fridges, and the slide on fertility) use household data. In each example, there are huge differences between countries on every level. Very few countries follow these lines exactly, but the lines show the general pattern of all countries over several decades. You can explore the actual plotted bubbles behind these lines at gapm.io/flinex.

What part of the line are you seeing? Many lines that are not straight can look straight if you zoom in enough—even a circle. This idea was inspired by Ellenberg (2014), *How Not to Be Wrong: The Power of Mathematical Thinking*. See gapm.io/fline.

Chapter Four: The Fear Instinct

Natural disasters. The numbers for the Nepal earthquake are from PDNA. Numbers for the 2003 heat wave in Europe are from UNISDR. All other disaster data is from EM-DAT. Nowadays, Bangladesh has a very cool flood-monitoring website; see http://www.ffwc.gov.bd. See gapm.io/tdis.

Child deaths from diarrhea. Our calculations of child deaths from diarrhea from contaminated drinking water are based on numbers from IHME[11] and WHO[4]. See gapm.io/tsan.

Plane accidents. The data on fatalities in recent years is from IATA and the data on passenger miles is from the UN agency that managed to reduce the number of accidents, see ICAO [1,2,3]. See gapm.io/ttranspa.

Deaths in wars. The figure of 65 million World War II deaths includes all deaths and comes from White[1,2]. The data sources for battle deaths (Correlates of War Project, Gleditsch, PRIO and UCDP[1]) include reported deaths of civilians and soldiers during battle, but not indirect deaths like those from starvation. Estimates of fatalities in Syria are from UCDP[2]. We strongly recommend watching this interactive data-driven documentary, which puts all known wars in perspective: www.fallen.io. To interactively compare fatalities in wars since 1990, go to http://ucdp .uu.se. See gapm.io/twar.

Fear of nuclear. The data on Fukushima is from the National Police Agency of Japan and Ichiseki (2013). According to police records, the Tōhoku earthquake and tsunami caused 15,894 confirmed deaths, and 2,546 people are still missing (as of December 2017). Tanigawa et al. (2012) concluded that 61 very old people in critical health conditions died during the hasty evacuation. About 1,600 further deaths were indirectly caused by other kinds of problems for mainly elderly evacuees, reports Ichiseki. According to Pew[1], in 2012, 76 percent of people in Japan believed that food from Fukushima was dangerous. The discussion of health investigations after Chernobyl is based on WHO[5]. Data about nuclear warheads is from the website Nuclear Notebook. See gapm.io/tnuc.

Chemophobia. Gordon Gribble (2013) tracks the origin of chemophobia back to the publication of *Silent Spring* (1962), by Rachel Carson, and chemical accidents in the decades that followed. He argues that the exaggerated and irrational fear of chemicals today leads to wrong usage of common resources. See gapm.io/ffea.

Refusing vaccination. In the US, 4 percent of parents think that vaccines are not important, according to Gallup[3]. In 2016, Larson et al. found that, across 67 countries, an average of 13 percent of people were skeptical about vaccination in general. There were huge variations between countries: from more than 35 percent in France and Bosnia and Herzegovina to 0 percent in Saudi Arabia and Bangladesh. In 1990, measles was the cause of 7 percent of all child deaths. Today, thanks to vaccination, it is only 1 percent. Deaths from measles mainly happen on Level 1 and Level 2, where children only recently started to get vaccinated; see IHME[7] and WHO[1]. See gapm.io/tvac.

DDT. Paul Hermann Müller won the Nobel Prize in Physiology and Medicine in 1948 for "his discovery of the high efficiency of DDT as a contact poison against several arthropods." Hungary was the first country to ban DDT, in 1968, followed by Sweden in 1969. The United States banned it three years later; see CDC[2]. An international treaty against various pesticides, including DDT, has since entered into force in 158 countries; see http://www.pops.int. Since the 1970s, CDC[4] and EPA have issued directives on how to avoid the dangers of DDT to humans. Today, the World Health Organization promotes the use of DDT to save lives in poor settings by killing malaria mosquitoes, within strict safety guidelines; see WHO[6, 7].

Terrorism. The data about fatalities from terrorism comes from the Global Terrorism Database; see GTD. The data on terror deaths per income level comes from Gapminder[3]. See Gallup[4] for the poll about fear of terrorism. See gapm.io/tter.

Alcohol deaths. Our calculations on deaths involving alcohol draw on IHME[9], NHTSA (2017), FBI, and BJS. See gapm.io/alcterex.

Risks of dying. The percentages we quote take the death tolls on Level 4 for the past ten years divided by the number of all deaths on Level 4 over that period, and are based on the following data sources: EM-DAT for natural disasters, IATA for plane crashes, IHME[10] for murders, UCDP[1] for wars, and GTD for terrorism. A more relevant risk calculation should not just divide by the number of all deaths, but rather should take into account exposure to the situations in which these kinds of deaths can occur. See gapm.io/ffear.

Comparing disasters. To compare different kinds of disaster deaths, see "Not All Deaths Are Equal: How Many Deaths Make a Natural Disaster Newsworthy?" online at OurWorldInData[8]. Gapminder is currently compiling data about the skewed media coverage of different kinds of deaths and different kinds of environmental problems. When ready, it will be published here: gapm.io/fndr.

Chapter Five: The Size Instinct

Nacala child deaths calculation. The births and population data used for these calculations is based on the Mozambique census of 1970, the Nacala hospital's own records, and UN-IGME of 2017.

Wrong proportions. The examples of proportions that people tend to overestimate come from Ipsos MORI[2,3], which reveal misconceptions across 33 countries. Paulos, *Innumeracy* (1988), is full of fascinating examples of disproportionality, asking, for example, how much the level of the Red Sea would rise if you added all the human blood in the world. See gapm.io/fsize.

Educated mothers and child survival. The discussion on how educated mothers lead to higher child survival is based on a study of data from 175 countries between 1970 and 2009, by Lozano, Murray et al. (2010). See gapm.io/tcare.

Saving lives. The list of the low-cost, high-impact interventions that save the most lives comes from UNICEF[2], which also set out the essential basic health care to which all citizens should have access before public health budgets start being spent on more advanced care.

4.2 million. The data on infant deaths in recent years comes from UN-IGME. The data on births and infant deaths in 1950 comes from UN-Pop[3].

Bears and axes. This striking comparison was brought to the public's awareness by a man named Hans Hansson. He wrote to his local newspaper about the absurd neglect of domestic violence against women and went on to start a network for men to help them break their violent behavior. Read an interview with him in English here: http://www.causeofdeathwoman .com/the-mens-network.

The Spanish flu. Crosby (1989), in his book *America's Forgotten Pandemic*, estimated that the Spanish flu caused 50 million deaths. The number is confirmed by Johnson and Mueller (2002) and CDC[1]. The world population in 1918 was 1.84 billion, which means this pandemic wiped out 2.7 percent of the entire global population.

TB and swine flu. The data on swine flu comes from WHO[17], and the data for TB from WHO[10,11]. See gapm.io/bswin.

Energy sources. The data comparing energy sources is from Smil, *Energy Transitions: Global and National Perspectives* (2016). Smil describes the slow transition away from fossil fuels and also debunks myths about food production, innovation, population, and mega-risks. See gapm .io/tene.

Future consumers. For an interactive visualization of the graphs on page 138, see gapm.io/incm. Two great books on this are *The Post-American*

World by Fareed Zakaria (2008) and *The World Is Flat* by Thomas L. Friedman (2005).

CO_2 **per capita.** The data on CO_2 per capita for China, the United States, Germany, and India comes from CDIAC. See gapm.io/tco2.

Chapter Six: The Generalization Instinct

Graph: Difference within Africa. For an interactive version of the graph on page 159, see gapm.io/edafr.

Contraception. The data comes from UNFPA[1] and UN-Pop[9]. See gapm.io/twmc.

Everything is made from chemicals. People with chemophobia divide the world into "natural" (safe) and "chemical" (industrial and harmful). The world's largest database of defined chemical compounds sees it differently. CAS contains 132 million organic and synthetic chemicals and their properties. It shows that toxicity is not related to who produces the compound. Cobratoxin (CAS registry number 12584-83-7), for example, which is produced by nature, paralyzes your nervous system until you can't breathe. See gapm.io/tind.

The Salhi family. See more about the Salhi family at gapm.io/dssah. If you think we have too few homes from Tunisia or elsewhere on gapm.io /dstun, feel free to contribute. Read more about how you can do it at: http://www.gapminder.org/dollar-street/about.

The recovery position. For more on the history of the recovery position, see Högberg and Bergström (1997) and Wikipedia[10].

Sudden infant death syndrome (SIDS). The conclusion that it was public health policy on the prone position that caused the increase in SIDS in Sweden is described by Högberg and Bergström (1997) and Gilbert et al. (2005). The report from Hong Kong is from Davies (1985).

Chapter Seven: The Destiny Instinct

The sense of superiority. For more on the sense of superiority over other groups, see Haidt, *The Righteous Mind: Why Good People Are Divided by Politics and Religion* (2012). See gapm.io/fdes.

Societies and cultures move. To see the World Health Chart in motion over 200 years, visit www.gapminder.org/whc and click Play.

Africa can catch up. The data for life expectancy for countries and regions comes from Gapminder[4]. Paul Collier writes in *The Bottom Billion* (2007) about the future prospects for the world's poorest people. Our rough estimate of people in extreme poverty close to conflicts is based on ODI (2015), preliminary results by Andreas Forø Tollefsen and Gudrun Østby of the number of people who live close to conflict worldwide (743 millions in 2016), and maps from WorldPop, IHME[6], FAO[4] and UCDP[2]. See the speed of improvement over the past decades here: gapm.io/edafr2.

Progress in China, Bangladesh, and Vietnam. *The Population Bomb*, by Paul and Anne Ehrlich (1968), contributed to a widespread idea that Asia and Africa would never be able to feed their growing populations. The data on deaths from famines is from EM-DAT. The Peace Research Institute Oslo (PRIO) produces maps of conflicts and poverty: gapm.io /mpoco. For global textile production, see gapm.io/tmante.

IMF forecasts. Our comments on the IMF's forecasting track record are based on the World Economic Outlook IMF[2]. See gapm.io/eecof.

Fertility in Iran. Professor Hossein Malek-Afzali, at Tehran University of Medical Science, was my host in Iran. He showed me the infertility clinic and taught me about Iran's family planning and sexual education programs. To compare Iran—the world champion in family planning— against other countries over time, see gapm.io/vm2.

Religions and babies. In most countries, a majority of the population belongs to one of the large religions, and this guides which chart the country shows up in. However, in many countries there is no clear majority. In Nigeria, for example, 49 percent of the population was Christian and 48 percent Muslim in 2010 according to our data on religion, Pew[2,3]. We have split 81 such countries into three separate bubbles in the relevant charts, using Pew[2] and USAID-DHS[2] to estimate each religious group's fertility rate, and roughly estimating each religious group's per capita income based on GDL[1,2], OECD[3] and other sources. See: gapm.io/ereltfr.

Asian values. In "Explaining Fertility Transitions" (1997), Karen Oppenheim Mason discusses changing family norms. Gender roles change quite fast in all cultures as people get richer and their way of living is modernized. In cultures with an emphasis on extended families, values may change a bit more slowly. See gapm.io/twmi.

Asian University for Women in Bangladesh. See http://www.auw.edu.bd.

Nature reserves. The data on protected nature is based on data from The World Database on Protected Areas (UNEP[5]), with the Protected Planet report (UNEP[6]) and IUCN[1, 2]. The trend for 1911–1990 comes from *Looking Ahead: The 50 Trends That Matter*; see Abouchakra et al. (2016). See Gapminder[5] for details.

Outdated chimpanzee questions. In the 1990s, students at Karolinska Institutet did not know that many European countries had worse health outcomes than many Asian countries. These are the results I show in my first TED talk: Rosling (2006). Thirteen years later, when we wanted to check whether people's knowledge had improved, we could not use the original questions, because these European countries had caught up, as shown in the animated chart here: gapm.io/vm3.

Cultural change in the United States and Sweden. The data on attitudes toward same-sex marriage in the United States is from Gallup[5].

Chapter Eight: The Single Perspective Instinct

Poll results from groups of professionals. For poll results for the groups of professionals mentioned here, and others, see gapm.io/rrs.

Expert forecasts. People with extraordinary expertise in one field score just as badly on our fact questions as everyone else. This wouldn't surprise Philip E. Tetlock and Dan Gardner, the authors of *Superforecasting* (2015). In this book they describe a systematic way to test people's ability to predict the future, and they find that one thing that can really impair good judgment is narrow expertise. They also describe the personality traits that often come with good judgment: humility, curiosity, and a willingness to learn from mistakes. You can practice your forecasting in their Good Judgment project: www.gjopen.com.

Lindau Nobel laureate meeting. This is a great annual gathering of brilliant young researchers who, thanks to this wonderful organization, get the chance to learn from Nobel laureates once a year. We are not criticizing that! We are just using their really low score on the vaccination question to make the case that expert knowledge doesn't guarantee general knowledge. Read more about the presentation on the Lindau website: gapm.io/xlindau64.

Plundered natural resources. For discussions about the commons and how to avoid exploitation, see *The Plundered Planet: Why We Must—*

and How We Can—Manage Nature for Global Prosperity, by Paul Collier (2010), and IUCN Red List[4].

Education needs electricity. For more on this, see UNDESA.

US health spending. The spending data comes from WHO[12]. The comparison between US spending and spending in other capitalist countries on Level 4 comes from OECD[1], a study named "Why Is Health Spending in the United States So High?" It concludes that costs in the US health-care system are higher across the board, but in particular costs of outpatient care and administration; and that this does not lead to better outcomes, because the system is not incentivizing doctors to spend time with the patients most in need of care. See gapm.io/theasp.

Democracy. Paul Collier's books are just as disturbing as they are fact-based. See his *Wars, Guns and Votes: Democracy in Dangerous Places* (2011) for more on how democracy can destabilize countries on Level 1 rather than make them safer. More disturbing problems with democracy are discussed in Fareed Zakaria's *The Future of Freedom: Illiberal Democracy at Home and Abroad*. We must remind ourselves of Winston Churchill's wise words: "No one pretends that democracy is perfect or all-wise. Indeed it has been said that democracy is the worst form of Government except for all those other forms that have been tried from time to time." See gapm.io/tgovd.

Fast economic growth and democracy. This discussion is based on economic growth data from IMF[1] and the Democracy Index 2016, from *The Economist*[2]. This index gives countries "democracy" ratings between 1 and 10, with the lowest score, 1.8, going to North Korea and the highest score, 9.93, to Norway. Here are the ten countries with the fastest economic growth over the past five years and their democracy scores (fastest first): Turkmenistan, 1.83; Ethiopia, 3.6; China, 3.14; Mongolia, 6.62; Ireland, 9.15; Uzbekistan, 1.95; Myanmar, 4.2; Laos, 2.37; Panama, 7.13; Georgia, 5.93. Only one of the ten fastest-growing economies scores well on democracy.

Chapter Nine: The Blame Instinct

Neglected diseases. For the list of diseases that are not profitable to the pharmaceutical industry, since the victims are almost entirely people living on Level 1, see WHO[15]. Until recently, Ebola was on this list.

Systems thinking. Peter Senge developed the idea of systems thinking within corporate organizations as a way of stopping people from blaming one another and helping them to understand the mechanisms that are causing problems. But his ideas apply to all kinds of human organizations where blaming individuals blocks understanding. See Senge, *The Fifth Discipline: The Art & Practice of the Learning Organization* (1990). See gapm.io/fblame.

UNICEF's low costs. UNICEF's streamlined logistics and supply chain are amazing. If you want to place a bid, you can see the supplies and services UNICEF is looking for right now at www.unicef.org/supply/index _25947.html. You can read more about its procurement process at UNICEF[5].

Why refugees don't fly. Sweden did not confiscate the boats of those smuggling refugees from Denmark during the Second World War—see the BBC documentary "How the Danish Jews Escaped the Holocaust." According to Goldberger (1987), 7,220 Danish Jews were saved by these boats. Today, EU Council[1] Directive 2002/90/EC defines "smuggler" as anyone facilitating illegal immigration, and an EU Council[2] framework decision allows "confiscation of the means of transport used to commit the offence." While the Geneva Conventions say that many of these refugees have the right to asylum, see UNHCR. See gapm.io/p16 and gapm.io/tpref.

CO$_2$ emissions. Researchers are trying to figure out how to adjust emissions quotas for changing population sizes; see Shengmin et al. (2011) and Raupach et al. (2014). See gapm.io/eco2a. For more on CO$_2$ emissions at different incomes, see gapm.io/tco2i.

Syphilis. If you think you are not living in the best of times, search for images of syphilis and you will soon feel blessed. We got the many names of this disgusting disease from Quétel (1990) via the University of Glasgow Library.

One billion people and Mao. One billion is a rounded-down approximation of the number of people whose lives were affected by Chairman Mao. In 1949, China's population was 0.55 billion. Mao ruled the country from 1949 until his death in 1976, during which time another 0.7 billion Chinese people were born, according to UN-Pop[1].

Falling birth rates and powerful leaders. This interactive chart shows how all countries' birth rates have fallen since 1800: gapm.io/vm4.

Abortion. The WHO guidelines on Access to Safe Abortion say: "Restriction in access to safe abortion services results in both unsafe abortions and unwanted births. Almost all deaths and morbidity from unsafe abortion occur in countries where abortion is severely restricted in law and/or in practice." See WHO[2].

Institutions. Institutions are best understood through the work performed by the people maintaining them. In their book *Poor Economics*, Banerjee and Duflo (2011) describe the very basic institutions needed to make the escape out of poverty easier. See gapm.io/tgovin.

The governmental employees who saved the world from Ebola. Dr. Mosoka Fallah was one of the Ebola contact tracers I had the honor of working with in Monrovia. Listen to his own words about the government's employees and their commitment to their society when it needed them most, and hear him describe how to maintain trust within the community while hunting the infection, in his TEDx Monrovia talk here: gapm.io/x1.

Thank you, industrialization. See the magic washing machine in action in this TED talk: gapm.io/vid1.

Chapter Ten: The Urgency Instinct

Konzo. To understand the lives of the villagers and their children suffering from konzo, watch the film by Thorkild Tylleskär (1995), recorded in the Bandundu Province, in present-day Democratic Republic of Congo: gapm.io/x2.

Now or never. Learn to defend yourself against common sales tricks in Robert Cialdini's *Influence* (2001).

The urgency instinct. See *Superforecasting*, by Tetlock and Gardner (2015), for more on how difficult it is for us to maintain "maybe," and therefore a reasonable range of options, in our heads.

The melting ice cap. The website Greenland Today shows the melting at the North Pole every day; see https://nsidc.org/greenland-today.

Fresh numbers for GDP and CO_2. The OECD regularly publishes data for its 35 rich member countries. As of December 2017, the most recent number for GDP growth is from six weeks ago. The most recent number for CO_2 emissions is from three years ago; see OECD[2]. For Sweden, CO_2 emissions data that is not older than three months can be found at the website for Sweden's System of Environmental and Economic Accounts; see SCB.

Climate refugees. Many studies claim to show that the number of refugees will increase dramatically because of climate change. The UK Government Office for Science study *Migration and Global Environmental Change* (Foresight, 2011) showed fundamental weaknesses in the common assumptions underlying these claims. First it found that most of the frequently quoted studies refer back to just two original sources, one estimating that climate change will create ten million refugees and the other anticipating 150 million refugees; see Box 1.2: "Existing estimates of 'numbers of environmental migrants' tend to be based on one or two sources." And second, it found that these original sources underestimate people living on Levels 1 and 2 and their ability to cope with change. Instead they describe migration as their only option in the face of climate change.

The bad habit of reducing all problems to one single problem—the climate—is called climate reductionism. To confront it is not to deny climate change. It is to have realistic expectations about how people will cope with it, bearing in mind the many examples in world history of humans adapting to new circumstances; see, for example, *The Big Ratchet*, by Ruth DeFries (2014).

For a fact-based picture of the global migration and refugee situation, see UNHCR Population Statistics here: http://popstats.unhcr.org/en/overview, and read Paul Collier's *Exodus* (2013), and Alexander Betts and Paul Collier's *Refuge* (2017).

Ebola. The WHO[13] lists all situation reports produced to track the Ebola pandemic since 2014. They still show suspect cases, and the CDC[3] continues to use the high estimates, which include suspected and unconfirmed cases.

The five global risks. For a fact-based view of a longer list of major risks, see *Global Catastrophes and Trends: The Next Fifty Years*, by Smil (2008). For those who find numbers calming, this is where you will find the big picture of the proportional risks and uncertainties of all kinds of possible fatal discontinuities. See gapm.io/furgr.

The risk of global pandemic. A small version of Spanish flu is more likely than a large one; see Smil (2008). While we should work against the obscene overuse of antibiotics in the meat industry—see WHO[14]—at the same time we must be careful not to make the mistake we made with DDT and become overprotective. Antibiotics could save even more lives if they were even less expensive. See gapm.io/tgerm.

The risk of financial collapse. During the past ten years, the external environment is volatile, with capital markets increasingly characterized by more extreme events, see Dobbs et al. in *No Ordinary Disruption* (2016). See also Hausmann (2015). See gapm.io/dysec.

The risk of World War III. In his book (2008), Smil was already discussing ten years ago how six unfolding trends of the new world order were slowly leading to intensified conflicts between parts of the world: Europe's place, Japan's decline, Islam's choice, Russia's way, China's rise, and the United States' retreat. See gapm.io/dysso.

The risk of climate change. The passage draws on *The Plundered Planet*, by Paul Collier (2010), the thinking of economist Elinor Ostrom and OurWorldInData[7]. See gapm.io/dysna.

The risk of extreme poverty. The passage draws on World Bank[26], ODI, PRIO, Paul Collier's *The Bottom Billion* (2007), and the BBC documentary "Don't Panic—End Poverty" (see Gapminder[11]). While extreme poverty has fallen, the number of extremely poor people living in conflict has been stable or even increased, based on preliminary data from PRIO. If current wars continue, soon the vast majority of extremely poor children will live behind military lines. This poses a cultural challenge to the international aid community; see the Stockholm Declaration (2016). See gapm.io/tepov.

Chapter Eleven: Factfulness in Practice

Diversified economies. MIT has produced a free-of-charge tool (https://atlas.media.mit.edu/en/) to help countries work out how best to diversify, given its existing industries and skills; see gapm.io/x4 or read Hausmann et al. (2013).

Teachers. Visit www.gapminder.org/teach to find our free teaching materials and join the community of teachers promoting a fact-based worldview in their classrooms.

Speling miskates. This typo is intentional, inspired by the fact that oriental rugs should always contain at least one deliberate mistake. At least one knot must always be wrong in every rug. It is to remind us that we are humans and we should not pretend we are perfect. Deliberately, we have no source behind this fact.

Constructive news. Here are two very different approaches for fixing the news problem: https://constructiveinstitute.org and https://www.wikitribune.com/.

Local ignorance and data. Don't miss Alan Smith's TEDx talk "Why you should love statistics" where he shows great examples of local misconceptions in the UK. Gapminder is starting to develop localized visualizations, like these about Stockholm. Every bubble represents a small area of the city. Push Play and see how 90 percent of areas are somewhere in the middle, and how most of Stockholm is getting richer and more educated, even as Stockholm political debate often discusses the people living at either extreme, because the differences are disturbingly large. See gapm.io/gswe1.

A Final Note

Free global development data. Open access to data and research made this book possible. In 1999, the World Bank produced, on a CD-ROM, the most comprehensive set of global statistics ever: "World Development Indicators." We uploaded the content to our website in our animated bubble graphs to make it easier for people to use. The World Bank got a bit angry, but our argument was that taxpayers had already paid for this official data to be collected; we were just making sure they could reach what they already owned. And we asked, "Don't you believe in free access to information in order for global market forces to work as they should?" In 2010, the World Bank decided to release all of its data for free (and thanked us for insisting). We presented at the ceremony for their new Open Data platform in May 2010, and since then the World Bank has become the main access point for reliable global statistics; see gapm.io/x6.

This was all possible thanks to Tim Berners-Lee and other early visionaries of the free internet. Sometime after he had invented the World Wide Web, Tim Berners-Lee contacted us, asking to borrow a slide show that showed how a web of linked data sources could flourish (using an image of pretty flowers). We share all of our content for free, so of course we said yes. Tim used this "flower-powerpoint" in his 2009 TED talk—see gapm.io/x6—to help people see the beauty of "The Next Web," and he uses Gapminder as an example of what happens when data from multiple sources come together; see Berners-Lee (2009). His vision is so bold, we have thus far seen only the early shoots!

Unfortunately, this book uses almost no data from the International Energy Agency (www.iea.org), which, together with OECD, still puts price tags on lots of taxpayers' data. That probably will—and has to—change soon, as energy statistics are way too important to remain so inaccessible.

SOURCES

Abouchakra, Rabih, Ibrahim Al Mannaee, and Mona Hammami Hijazi. *Looking Ahead: The 50 Trends That Matter.* Chart, page 274. Bloomington, IN: Xlibris, 2016.

Allansson, Marie, Erik Melander, and Lotta Themnér. "Organized violence, 1989–2016." *Journal of Peace Research* 54, no. 4 (2017).

Amnesty. *Death Penalty: Data counting abolitionists for all crimes.* 2007–2016. Accessed November 3, 2017. gapm.io/xamndp17.

Ariely, Dan. *The Honest Truth About Dishonesty: How We Lie to Everyone, Especially Ourselves.* New York: Harper, 2012.

———. *Predictably Irrational: The Hidden Forces That Shape Our Decisions.* New York: Harper, 2008.

———. *The Upside of Irrationality, The Unexpected Benefits of Defying Logic at Work and at Home.* New York: Harper, 2010.

ATAA. Air Transport Association of America. *The Annual Reports of the U.S. Scheduled Airline Industry,* 1940 to 1991. Earlier editions were called *Little known facts about the scheduled air transport industry* and *AIR Transport Facts and Figures.* Accessed November 2017. http://airlines.org.

Banerjee, Abhijit Vinayak, and Esther Duflo. *Poor Economics: A Radical Rethinking of the Way to Fight Global Poverty.* New York: PublicAffairs, 2011.

Barro-Lee. Educational Attainment Dataset v 2.1. Updated February 4, 2016. See Barro and Lee (2013). Accessed November 7, 2017. http://www.barrolee.com. gapm.io/xbl17.

Barro, Robert J., and Jong-Wha Lee. "A New Data Set of Educational Attainment in the World, 1950–2010." *Journal of Development Economics* 104 (2013): 184–98.

BBC. Producer Farhana Haider. "How the Danish Jews escaped the Holocaust." Witness, *BBC, Magazine, October 14*, 2015. gapm.io/xbbcesc17.

Berners-Lee, Tim. "The next web." Filmed February 2009 in Long Beach, CA. TED video, 16:23. gapm.io/x-tim-b-l-ted. https://www.ted.com/talks/tim_berners_lee_on_the_next_web.

Betts, Alexander, and Paul Collier. *Refuge: Rethinking Refugee Policy in a Changing World.* New York: Oxford University Press, 2017.

Biraben, Jean-Noel. "An Essay Concerning Mankind's Evolution." *Population.* Selected Papers. Table 2. December 1980. As cited in US Census Bureau. gapm.io/xuscbbir.

BJS (Bureau of Justice Statistics). Rand, M. R., et al. *"Alcohol and Crime: Data from 2002 to 2008".* Washington, DC: Bureau of Justice Statistics, Office of Justice Programs, US Department of Justice, 2010. Page last revised on July 28, 2010. Accessed December 21, 2017. https://www.bjs.gov/content/acf/ac_conclusion.cfm.

Bongaarts, John, and Rodolfo A. Bulatao. "Beyond Six Billion: Forecasting the World's Population." National Research Council. Panel on Population Projections. Committee on Population, Commission on Behavioral and Social Sciences and Education. Washington, D.C. 2000. National Academy Press. https://www.nap.edu/read/9828/chapter/4#38.

Bourguignon, François, and Christian Morrisson. "Inequality Among World Citizens: 1820–1992." *American Economic Review* 92, no. 4 (September 2002): 727–44.

Bryant, John. "Theories of Fertility Decline and the Evidence from Development Indicators." *Population and Development Review* 33, no. 1 (March 2007): 101–27.

BTS[1]. (US Bureau of Transportation Statistics). US Air Carrier Safety Data. Total fatalities. National Transportation Statistics. Table 2-9. Accessed November 24, 2017. gapm.io/xbtsafat.

BTS[2]. Revenue Passenger-miles (the number of passengers and the distance flown in thousands (000)). T-100 Segment data. Accessed November 4, 2017. gapm.io/xbtspass.

Caldwell, J. C. "Three Fertility Compromises and Two Transitions." *Population Research and Policy Review* 27, no. 4 (2008): 427–46. gapm.io /xcaltfrt.

Carson, Rachel. *Silent Spring*. Boston: Houghton Mifflin, 1962.

CAS. Database Counter. American Chemical Society, 2017. Accessed December 3, 2017. gapm.io/xcas17.

CDC[1]. (Center for Disease Control and Prevention). Taubenberger, Jeffery K., and David M. Morens. "1918 Influenza: The Mother of All Pandemics." *Emerging Infectious Diseases* 12, no. 1 (January 2006): 15–22. gapm.io/xcdcsflu17.

CDC[2]. "Organochlorine Pesticides Overview" Dichlorodiphenyltrichloroethane (DDT). National Biomonitoring Program.

CDC[3]. "Ebola Outbreak in West Africa—Reported Cases Graphs." Centers for Disease Control and Prevention, 2014. gapm.io/xcdceb17.

CDC[4]. Toxicological Profile for DDT, DDE and DDD https://www .atsdr.cdc.gov/toxprofiles/tp.asp?id=81&tid=20.

CDIAC. *"Global, Regional, and National Fossil-Fuel CO2 Emissions"* Boden, T.A., G. Marland, and R.J. Andres. 2017. Carbon Dioxide Information Analysis Center, Oak Ridge National Laboratory, U.S. Department of Energy, Oak Ridge, Tenn., U.S.A. DOI 10.3334/CDIAC /00001_V2017. gapm.io/xcdiac.

CETAD (Centro de Estudos Tributários e Aduaneiros). "Distribuição da Renda por Centis Ano MARÇO 2017." Ministério da Fazenda (Brazil), 2017. gapm.io/xbra17.

Cialdini, Robert B. *Influence: How and Why People Agree to Things*. Boston, MA: Allyn and Bacon, 2001.

College Board. SAT Total Group Profile Report, 2016. gapm.io/xsat17.

Collier, Paul. *The Bottom Billion: Why the Poorest Countries Are Failing and What Can Be Done About It*. New York: Oxford University Press, 2007.

———. *Exodus: How Migration Is Changing Our World*. New York: Oxford University Press, 2013.

———. *The Plundered Planet: Why We Must—and How We Can—Manage Nature for Global Prosperity*. New York: Oxford University Press, 2010.

———. *Wars, Guns and Votes: Democracy in Dangerous Places*. New York: Random House, 2011.

Correlates of War Project. COW Dataset v4.0. Based on Sarkees, Meredith Reid, and Frank Wayman (2010). Dataset updated 2011. Accessed Dec 3, 2017. http://www.correlatesofwar.org/data-sets/COW-war.

Countdown to 2030. *Reproductive, Maternal, Newborn, Child, and Adolescent Health and Nutrition.* Data produced by Aluisio Barros and Cesar Victora at Federal University of Pelotas, Brazil, 2017. http://countdown2030.org/.

Crosby, Alfred W. *America's Forgotten Pandemic.* Cambridge, UK: Cambridge University Press, 1989.

Cummins, Denise. "Why the Gender Difference on SAT Math Doesn't Matter." *Good Thinking* blog, *Psychology Today.* March 17, 2014.

Davies, D.P. (1985). "Cot death in Hong Kong: a rare problem?" Lancet 1985 Dec 14;2(8468):1346-9. https://www.ncbi.nlm.nih.gov/pubmed/2866397.

DeFries, Ruth. *The Big Ratchet: How Humanity Thrives in the Face of Natural Crisis.* New York: Basic Books, 2014.

Diamond, Jared. *The World Until Yesterday: What Can We Learn from Traditional Societies?* London: Viking, 2012.

Dobbs, Richard, James Manyika, and Jonathan Woetzel. *No Ordinary Disruption: The Four Global Forces Breaking All the Trends.* New York: PublicAffairs, 2016.

Dollar Street. Free photos under Creative Commons BY 4.0. By Gapminder, Anna Rosling Rönnlund. 2017. www.dollar-street.org.

Ehrlich, Paul R., and Anne Ehrlich. *The Population Bomb.* New York: Ballantine, 1968.

EIA. US Energy Information Administration. "Annual passenger travel tends to increase with income." International Energy Outlook, Bureau of Transportation Statistics, National Transportation Statistics, 2016.

Ellenberg, Jordan. *How Not to Be Wrong: The Power of Mathematical Thinking.* New York: Penguin, 2014.

Elsevier. Reller, Tom. "Elsevier Publishing—A Look at the Numbers, and More." Posted March 22, 2016. Accessed November 26, 2017. https://www.elsevier.com/connect/elsevier-publishing-a-look-at-the-numbers-and-more.

EM-DAT. Centre for Research on the Epidemiology of Disasters (CRED). The International Disaster Database. Debarati Guha-Sapir, Université catholique de Louvain. Accessed November 5, 2017. www.emdat.be.

Encuesta Nacional de Ingresos y Gastos de los Hogares (ENIGH) 2016. Tabulados básicos. 2017. Table 2.3, 2016.

EPA (US Environmental Protection Agency). Environment Program, Pesticide information. gapm.io/xepa17.

EU Council[1]. Council Directive 2002/90/EC of 28 November 2002 "defining the facilitation of unauthorised entry, transit and residence." November, 2002. gapm.io/xeuc90.

EU Council[2]. Council Directive 2001/51/EC of 28 June 2001 "supplementing the provisions of Article 26 of the Convention implementing the Schengen Agreement of 14 June 1985." June 2001. gapm.io/xeuc51.

FAO[1]. (Food and Agriculture Organization of the United Nations). "Food Insecurity in the World 2006". 2006. gapm.io/faoh2006.

FAO[2]. *The State of World Fisheries and Aquaculture 2016: Contributing to Food Security and Nutrition for All.* Rome: FAO, 2016. Accessed November 29, 2017. http://www.fao.org/3/a-i5555e.pdf. gapm.io/xfaofi.

FAO[3]. "Statistics—Food security indicators." Last modified October 31, 2017. Accessed November 29, 2017. gapm.io/xfaofsec.

FAO[4]. FAOSTAT World Total, Yield: Cereals, Total, 1961–2014. Last modified May 17, 2017. Accessed November 29, 2017. gapm.io/xcer.

FAO[5]. "State of the World's Land and Water Resources for Food and Agriculture." SOLAW, FAO, Maps, 2011. gapm.io/xfaowl17.

FBI. Uniform Crime Reporting Statistics. *Crime in the United States.* All reported violent crimes and property crimes combined. Accessed October 12, 2017. gapm.io/xfbiu17.

Foresight. *Migration and Global Environmental Change. Final Project Report.* London: Government Office for Science, 2011. gapm.io/xcli17.

FRD. Ofcansky, Thomas P., Laverle Bennette Berry, and Library of Congress Federal Research Division. *Ethiopia: A Country Study.* Washington, DC: Federal Research Division, Library of Congress, 1993. gapm.io/xfdi.

Friedman, Thomas L. *The World Is Flat: A Brief History of the Twenty-first Century.* New York: Farrar, Straus & Giroux, 2005.

Gallup[1]. McCarthy, Justin. "More Americans Say Crime Is Rising in U.S." Gallup News, October 22, 2015. Accessed December 1, 2017. http://news.gallup.com/poll/186308/americans-say-crime-rising.aspx.

Gallup[2]. Brewer, Geoffrey. "Snakes Top List of Americans' Fears." Gallup News, March 19, 2001. Accessed December 17, 2017. http://news.gallup.com/poll/1891/snakes-top-list-americans-fears.aspx.

Gallup[3]. Newport, Frank. "In U.S., Percentage Saying Vaccines Are Vital Dips Slightly." Gallup News, March 6, 2015. gapm.io/xgalvac17.

Gallup[4]. "Concern About Being Victim of Terrorism." U.S. polls, 1995–2017. Gallup News, December 2017. gapm.io/xgal17.

Gallup[5]. McCarthy, Justin. "U.S. Support for Gay Marriage Edges to New High." Gallup News, May 3–7, 2017. gapm.io/xgalga.

Gapminder[1]. Regions, dividing the world into four regions with equal areas. gapm.io/ireg.

Gapminder[2]. GDP per capita—v25. Mainly Maddison data extended by Mattias Lindgren and modified by Ola Rosling to align with World Bank GDP per capita constant PPP 2011, with IMF forecasts from WEO 2017. gapm.io/dgdppc.

Gapminder[3]. Four income levels—v1. gapm.io/elev.

Gapminder[4]. Life expectancy v9, based on IHME-GBD 2016, UN Population and Mortality.org. Main work by Mattias Lindgren. gapm.io/ilex.

Gapminder[5]. Protected nature—v1—based on World Database on Protected Areas (WDPA), UK-IUCN, UNEP-WCMC. gapm.io/natprot.

Gapminder[6]. Child mortality rate—v10. Based on UN-IGME. Downloaded November 10, 2017, gapm.io/itfr.

Gapminder[7]. Total fertility rate—v12. gapm.io/dtfr.

Gapminder[8]. Income mountains—v3. Accessed November 2, 2017. gapm.io/incm.

Gapminder[9]. Extreme poverty rate—v1, rough guestimation of extreme poverty rates of all countries for the period 1800 to 2040, based on the Gapminder Income Mountains data set. gapm.io/depov.

Gapminder[10]. Household per capita income—v1. gapm.io/ihhinc.

Gapminder[11]. "Don't Panic—End Poverty." BBC documentary featuring Hans Rosling. Directed by Dan Hillman. Wingspan Productions, September 2015.

Gapminder[12]. Legal slavery data—v1. gapm.io/islav.

Gapminder[13]. HIV,newly infected—v2. Historic prevalence estimates before 1990 by Linus Bengtsson and Ziad El-Khatib. gapm.io/dhivnew.

Gapminder[14]. Death penalty abolishment—v1. gapm.io/ideat.

Gapminder[15]. Countries ban leaded gasoline—v1. gapm.io/ibanlead.

Gapminder[16]. Air plane fatalities—v1. Indicator Population—v5—all countries—1800–2100, based on UN WPP 2017 and mainly Maddison before 1950. gapm.io/dpland.

Gapminder[17]. Population—v5—all countries—1800–2100, based on UN-Pop WPP 2017 and mainly Maddison before 1950. gapm.io/dpop.

Gapminder[18]. Undernourishment—v1. gapm.io/dundern.

Gapminder[19]. Feature films—v1. gapm.io/dcultf.

Gapminder[20]. Women suffrage—v1—based primarily on Wikipedia page about women suffrage. gapm.io/dwomsuff.

Gapminder[21]. Literacy rate—v1—based on UIS and van Zanden. gapm. io/dliterae.

Gapminder[22]. Internet users—v1. gapm.io/dintus.

Gapminder[23]. Children with some vaccination—v1—based on WHO. gapm.io/dsvacc.

Gapminder[24]. Playable guitars per capita (very rough estimates)—v1. gapm.io/dguitars.

Gapminder[25]. Maternal mortality—v2. gapm.io/dmamo.

Gapminder[26]. "Factpods on Ebola." 1–15. gapm.io/fpebo.

Gapminder[27]. Poll results from events. gapm.io/rrs.

Gapminder[28]. How good are the UN population forecasts? gapm.io/ mmpopfut.

Gapminder[29]. The Inevitable Fill-Up. gapm.io/mmfu.

Gapminder[30]. Family size by income level. gapm.io/efinc.

Gapminder[31]. Protected Nature—v1. gapm.io/protnat.

Gapminder[32]. Hans Rosling. "Swine flu alert! News/Death ratio: 8176." Video. May 8, 2009. gapm.io/sftbn.

Gapminder[33]. Average age at first marriage. gapm.io/fmarr.

Gapminder[34]. World Health Chart. www.gapminder.org/whc.

Gapminder[35]. Differences within Africa. gapm.io/eafrdif.

Gapminder[36]. Monitored species. gapm.io/tnwlm.

Gapminder[37]. Food production. gapm.io/tfood.

Gapminder[38]. War deaths. gapm.io/twar.

Gapminder[39]. Textile. gapm.io/ttextile.

Gapminder[40]. Protected nature. gapm.io/protnat.

Gapminder[41]. "Why Boat Refugees Don't Fly!" gapm.io/p16.

Gapminder[42]. Child labour. gapm.io/dchlab.

Gapminder[43]. Gapminder Factfulness Poster, v3.1. Free Teaching Material, License CC BY. 4.0. 2017. gapm.io/fposter.

Gapminder[44]. Length of schooling. gapm.io/dsclex.

Gapminder[45]. Recreation spending by income level. gapm.io/tcrecr.

Gapminder[46]. Caries. gapm.io/dcaries.

Gapminder[47]. Fertility rates by income quintile. gapm.io/dtfrq.

Gapminder[48]. Road accidents. gapm.io/droada.

Gapminder[49]. Child drownings by income level. gapm.io/ddrown.

Gapminder[50]. Travel distance. gapm.io/ttravel.

Gapminder[51]. CO2 emissions. gapm.io/tco2.

Gapminder[52]. Natural disasters. gapm.io/tndis.

Gapminder[53]. Fertility rate and income by religion. gapm.io/dtfrr.

GDL[1]. (Global Data Lab). Area data initiated by Jeroen Smits. https://globaldatalab.org/areadata.

GDL[2]. IWI International Wealth Index. https://globaldatalab.org/iwi.

Gilbert et al. (2005). "Infant sleeping position and the sudden infant death syndrome: systematic review of observational studies and historical review of recommendations from 1940 to 2002" Ruth Gilbert, Georgia Salanti, Melissa Harden, Sarah. See International Journal of Epidemiology, Volume 34, Issue 4, 1 August 2005, Pages 874–887. https://doi.org/10.1093/ije/dyi088.

Gilovich, Thomas. *How We Know What Isn't So*. New York: Macmillan, 1991.

Gleditsch, Nils Petter. Norwegian: *Mot en mer fredelig verden?* [*Towards a more peaceful world?*]. Oslo: Pax, 2016. Figure 1.4. gapm.io/xnpgfred.

Gleditsch, Nils Petter, and Bethany Lacina. "Monitoring trends in global combat: A new dataset of battle deaths." *European Journal of Population* 21, nos. 2–3 (2005): 145–66. gapm.io/xbat.

Goldberger, Leo. *The Rescue of the Danish Jews: Moral Courage Under Stress*. New York: New York University Press, 1987.

Good Judgment Project. www.gjopen.com.

Gottschall, Jonathan. *The Storytelling Animal: How Stories Make Us Human*. Boston and New York: Houghton Mifflin Harcourt, 2012.

Gribble, Gordon W. "Food chemistry and chemophobia." *Food Security* 5, no. 1 (February 2013). gapm.io/xfosec.

GSMA. *The Mobile Economy 2017*. GSM Association, 2017. gapm.io/xgsmame.

GTD. Global Terrorism Database 2017. Accessed December 2, 2017. gapm.io/xgtdb17.

GTF. "The Global Tracking Framework measures the population with access to electricity in both rural and urban areas from 1990-2014." The World Bank & the International Energy Agency. Global Tracking Framework. Accessed November 29, 2017. http://gtf.esmap.org/results.

Gurven, Michael, and Hillard Kaplan. "Longevity Among Hunter-Gatherers: A Cross-Cultural Examination." *Population and Development Review* 33, no. 2 (2007): 321–65. gapm.io/xhun.

Haidt, Jonathan. *The Happiness Hypothesis: Finding Modern Truth in Ancient Wisdom*. New York: Basic Books, 2006.

———. *The Righteous Mind: Why Good People Are Divided by Politics and Religion*. New York: Pantheon, 2012.

Hausmann, Ricardo. "How Should We Prevent the Next Financial Crisis?" The Growth Lab, Harvard University, 2015. gapm.io/xecc.

Hausmann, Ricardo, Cesar A. Hidalgo, et al. *Atlas of Economic Complexity: Mapping Paths to Prosperity*, 2nd ed. Cambridge, MA: MIT Press, 2013. Accessed November 10, 2017. gapm.io/xatl17.

Hellebrandt, Tomas, and Paulo Mauro. *The Future of Worldwide Income Distribution*. Peterson Institute for International Economics Working Paper 15-7, April 2015. Accessed November 3, 2017. gapm.io/xpiie17.

HMD (Human Mortality Database). University of California, Berkeley and Max Planck Institute for Demographic Research. Downloaded September 2012. Available at www.mortality.org or www.humanmortality.de.

Högberg, Ulf, and Erik Bergström. "Läkarråd ökade risken för plötslig spädbarnsdöd" ["Physicians' advice increased the risk of sudden infant death syndrome"]. *Läkartidningen* 94, no. 48 (1997). gapm.io/xuhsids.

IATA (International Air Transport Association). "Accident Overview." Table. Fact Sheet Safety. December 2017. gapm.io/xiatas.

ICAO[1] (International Civil Aviation Organization). Convention on International Civil Aviation. Chicago, December 7, 1944. gapm.io/xchicc.

ICAO[2]. Aircraft Accident and Incident Investigation. Convention on International Civil Aviation, Annex 13. International Standards and Recommended Practices, 1955. gapm.io/xchi13.

ICAO[3]. Global Key Figures. Revenue Passenger-Kilometres. Air Transport Monitor. 2017. https://www.icao.int/sustainability/Pages/Air-Traffic-Monitor.aspx.

Ichiseki, Hajime. "Features of disaster-related deaths after the Great East Japan Earthquake." *Lancet* 381, no. 9862 (January 19, 2013): 204. gapm.io/xjap.

ICP. "Purchasing Power Parity $ 2011." International Comparison Program. gapm.io/x-icpp.

IHME[1] (Institute for Health Metrics and Evaluation). Data Life Expectancy. Global Burden of Disease Study 2016. Institute for Health Metrics and Evaluation, University of Washington, Seattle, September 2017. Accessed October 7, 2017. gapm.io/xihlex.

IHME[2]. "Global Educational Attainment 1970–2015." Accessed May 10, 2017. gapm.io/xihedu.

IHME[3]. "Road injuries as a percentage of all disability." GDB Compare. gapm.io/x-ihaj.

IHME[4]. "Drowning as a percentage of all death ages 5–14, by four development levels." GDB Compare. http://ihmeuw.org/49kq.

IHME[5]. "Drowning, share of all child deaths in ages 5–14, comparing Sweden with average for all highly developed countries." GBD Compare. http://ihmeuw.org/49ks.

IHME[6]. "Local Burden of Disease—Under-5 mortality." Accessed November 29, 2017. gapm.io/xih5mr.

IHME[7]. "Measles." GBD Compare. Institute for Health Metrics and Evaluation, University of Washington, 2016. gapm.io/xihels.

IHME[8]. "All causes of death" GBD Compare. Institute for Health Metrics and Evaluation, University of Washington, 2016. http://ihmeuw.org/49p3.

IHME[9]. "Transport injuries." GBD Compare. Institute for Health Metrics and Evaluation, University of Washington, 2016. http://ihmeuw.org/49pa.

IHME[10]. "Interpersonal violence." GDP Compare. Institute for Health Metrics and Evaluation, University of Washington, 2016. http://ihmeuw.org/49pc.

IHME[11]. Data for deaths under age 5 in 2016, attributable to risk factor unsafe water source, from IHME GBD 2016. Accessed December 12, 2017. http://ihmeuw.org/49xs.

ILO[1] (International Labour Organization). Forced Labour Convention, 1930 (No. 29) (C.29). Accessed December 2, 2017. gapm.io/xiloflc.

ILO[2]. Abolition of Forced Labour Convention, 1957 (No. 105) (C.105). Accessed December 2, 2017. gapm.io/xilola.

ILO[3]. Country baselines: Turkmenistan. gapm.io/xiloturkm.

ILO[4]. Country baselines: Uzbekistan. gapm.io/xilouzb.

ILO[5]. Country baselines: North Korea. gapm.io/xilonkorea.

ILO[6]. Convention No. 182 on the worst forms of child labour, 1999. gapm.io/xilo182.

ILO[7]. IPEC (Yacouba Diallo, Alex Etienne, and Farhad Mehran). "Global child labour trends 2008 to 2012." International Programme on the Elimination of Child Labour (IPEC). Geneva: ILO, 2013. gapm.io/xiloi.

ILO[8]. IPEC. Children in employment, child labour and hazardous work, 5–17 years age group, 2000–2012. Page 3, Table 1. International Labour Office; ILO International Programme on the Elimination of Child Labour (IPEC). gapm.io/xiloipe.

ILO[9]. "Programme on the Elimination of Child Labour, World (1950–1995)." International Labour Organization Programme on Estimates and Projections on the Elimination of Child Labour (ILO-EPEAP). Kaushik Basu, 1999. Via OurWorldInData.org/child-labor.

ILO[10]. Living Standard Measurement Survey., LABORSTA Labour Statistics Database. International Labour Organization. gapm.io/xilohhs.

ILMC (International Lead Management Center). Lead in Gasoline Phase-Out Report Card, 1990s. International Lead Zinc Research Organization (ILZRO), supported by the International Lead Association (ILA). Accessed October 12, 2017. http://www.ilmc.org/rptcard.pdf.

IMF[1] (International Monetary Fund). GDP per capita, constant prices with forecasts to 2022. World Economic Outlook 2017, October edition. Accessed November 2, 2017. gapm.io/ximfw.

IMF[2]. Archive. World Economic Outlook Database, previous years. gapm.io/ximfwp.

International Rhino Foundation. "Between 5,042–5,455 individuals in the wild—Population slowly increasing." Black Rhino. November 5, 2017. https://rhinos.org.

IMDb. Internet Movie Database. Search results for feature films filtered by year. gapm.io/ximdbf.

India Census 2011. "State of Literacy." Office of the Registrar General & Census Commissioner, India. 2011. gapm.io/xindc.

ISC (Internet System Consortium). "Internet host count history." gapm.io/xitho.

IPCC[1] (Intergovernmental Panel on Climate Change). Fifth Assessment Report (AR5) Authors and Review Editors. May 27, 2014. gapm.io/xipcca.

IPCC[2]. Fifth Assessment Report (AR5)—Climate Change 2014: Climate Change 2014 Synthesis Report, page 10: "Surface temperature is projected to rise over the 21st century under all assessed emission scenarios." Accessed April 10, 2017. gapm.io/xipcc.

Ipsos MORI[1]. Online polls for Gapminder in 12 countries, August 2017. gapm.io/gt17re.

Ipsos MORI[2]. "Perils of Perception 2015." Ipsos MORI, December 2, 2015. gapm.io/xip15.

Ipsos MORI[3]. "Perils of Perception 2016," Ipsos MORI, December 14, 2016. gapm.io/xip16.

IPUMS. Integrated Public Use Microdata Series International. Version 6.3. gapm.io/xipums.

ISRC. "International Standard Recording Code." Managed by International ISRC Agency. http://isrc.ifpi.org/en/faq.

ITOPF (International Tanker Owners Pollution Federation). "Oil tanker spill statistics 2016." Page 4. Published February 2017. Accessed September 20, 2017. http://www.itopf.com/fileadmin/data/Photos/Publications/Oil_Spill_Stats_2016_low.pdf.

ITRPV. "International Technology Roadmap for Photovoltaic." Workshop at Intersolar Europe, Munich, June 1, 2017. Graph on slide 6. gapm.io/xitrpv.

ITU[1] (International Telecommunication Union). "Mobile cellular subscriptions." World Telecommunication/ICT Development Report and Database. gapm.io/xitumob.

ITU[2]. "ICT Facts and Figures 2017." Individuals using the Internet. Accessed November 27, 2017. gapm.io/xituintern.

IUCN[1] (International Union for Conservation of Nature). Protected Area (Definition 2008). gapm.io/xprarde.

IUCN[2]. Categories of protected areas. gapm.io/x-protareacat.

IUCN[3]. Green, Michael John Beverley, ed. IUCN Directory of South Asian Protected Areas. IUCN, 1990.

IUCN Red List[1]. Goodrich, J., et al., "Panthera tigris (Tiger)." IUCN Red List of Threatened Species 2015: e.T15955A50659951. Accessed December 7, 2017. gapm.io/xiucnr1.

IUCN Red List[2]. Swaisgood, R., D. Wang, and F. Wei. Ailuropoda melanoleuca (Giant Panda) (errata version published in 2016). IUCN Red List of Threatened Species 2016: e.T712A121745669. Accessed December 7, 2017. http://dx.doi.org/10.2305/IUCN.UK.2016-2.RLTS.T712A45033386.en.

IUCN Red List[3]. Emslie, R. "Diceros bicornis (Black Rhinoceros, Hook-lipped Rhinoceros)." IUCN Red List of Threatened Species 2012: e.T6557A16980917. Accessed December 7, 2017. http://dx.doi.org/10.2305/IUCN.UK.2012.RLTS.T6557A16980917.en.

IUCN Red List[4]. IUCN. "Table 1: Numbers of threatened species by major groups of organisms (1996–2017)." Last modified September 14, 2017. gapm.io/xiucnr4.

Jacobson, Jodi L. "Environmental Refugees: A Yardstick of Habitability." Worldwatch Paper 86. Worldwatch Institute, 1988.

Jinha, A. E. "Article 50 million: an estimate of the number of scholarly articles in existence." *Learned Publishing* 23, no. 10 (2010): 258–63. DOI: 10.1087/20100308. gapm.io/xjinha.

Johnson, N. P., and J. Mueller. "Updating the accounts: global mortality of the 1918–1920 'Spanish' influenza pandemic." *Bulletin of the History of Medicine* 76, no. 1 (Spring 2002): 105–15.

Kahneman, Daniel. *Thinking, Fast and Slow.* New York: Farrar, Straus & Giroux, 2011.

Keilman, Nico. "Data quality and accuracy of United Nations population projections, 1950–95." *Population Studies* 55, no. 2 (2001): 149–64. Posted December 9, 2010. gapm.io/xpaccur.

Klein Goldewijk, Kees. "Total SO2 Emissions." Utrecht University. Based on Paddy (http://cdiac.ornl.gov). May 18, 2013. gapm.io/x-so2em.

Klepac, Petra, et al. "Towards the endgame and beyond: complexities and challenges for the elimination of infectious diseases." Figure 1. *Philosophical Transactions of the Royal Society B*, June 24, 2013. DOI: 10.1098/rstb.2012.0137. http://rstb.royalsocietypublishing.org/content/368/1623/20120137.

Lafond, F., et al. "How well do experience curves predict technological progress? A method for making distributional forecasts." Navigant Research. 2017. https://arxiv.org/pdf/1703.05979.pdf.

Larson, Heidi J., et al. "The State of Vaccine Confidence 2016: Global Insights Through a 67-Country Survey." *EBioMedicine* 12 (October 2016): 295–301. Posted September 13, 2016. DOI: 10.1016/j.ebiom.2016.08.042. gapm.io/xvacnf.

Lindgren, Mattias. "Gapminder's long historic time series." published from 2006 to 2016. gapm.io/histdata.

Livi-Bacci, Massimo. *A Concise History of World Population*, 2nd. ed. Page 22. Maiden, MA: Blackwell, 1989.

Lozano, Rafael, Krycia Cowling, Emmanuela Gakidou, and Christopher J. L. Murray. "Increased educational attainment and its effect on child mortality in 175 countries between 1970 and 2009: a systematic analysis." *Lancet*

376, no. 9745 (September 2010): 959–74. DIO: 10.1016/S0140-6736 (10) 61257-3. gapm.io/xedux.

Maddison[1]. Maddison project maintaining data from Angus Maddison. GDP per capita estimates, via CLIO Infra. Updated by Jutta Bolt and Jan Luiten van Zanden, et al. Accessed December 3, 2017. https://www .clio-infra.eu/Indicators/GDPperCapita.html.

Maddison[2]. Maddison project via CLIO Infra. Filipa Ribeiro da Silva's version revised by Jonathan Fink-Jensen, updated April 29, 2015. https:// www.clio-infra.eu/Indicators/TotalPopulation.html.

Magnus & Pia. Mino's parents.

McEvedy, Colin, and Richard Jones. *Atlas of World Population History*. New York: Facts on File, 1978. As cited in US Census Bureau. gapm.io/x -pophist.

Mischel, Walter. *The Marshmallow Test: Mastering Self-control*. New York: Little, Brown, 2014.

Music Trades. "The Annual Census of the Music Industries." 2016. http:// www.musictrades.com/census.html.

Myrskylä, M., H. P. Kohler, and F. Billari. "Advances in Development Reverse Fertility Declines." *Nature* 460, No. 6 (2009): 741–43. DOI: 10.1038/nature 08230.

National Biomonitoring Program. Centers for Disease Control and Prevention Organochlorine Pesticides Overview. gapm.io/xpes.

National Police Agency of Japan. *Damage Situation and Police Countermeasures Associated with 2011 Tohoku District - Off the Pacific Ocean Earthquake September 8, 2017*. Emergency Disaster Countermeasures Headquarters. gapm.io/xjapan.

NCI[1] (National Cancer Institute). "Trends in relative survival rates for all childhood cancers, age <20, all races, both sexes SEER (9 areas), 1975–94." Figure 10, p. 9, in L. A. G. Ries, M. A. Smith, et al., eds., "Cancer Incidence and Survival Among Children and Adolescents: United States SEER Program 1975–1995." National Cancer Institute, SEER Program. NIH. Pub. No. 99-4649. Bethesda, MD: 1999. gapm.io/xccs17.

NCI[2]. Childhood cancer rates calculated using the Incidence SEER18 Research Database, November 2016 submission (Katrina/Rita Population Adjustment). https://www.cancer.gov/types/childhood-cancers /child-adolescent-cancers-fact-sheet#r4.

NHTSA (National Highway Traffic Safety Administration). "Alcohol-Impaired Driving from the Traffic Safety Facts, 2016 Data." Table 1. October 2017. gapm.io/xalc.

Nobel Prize in Physiology or Medicine 1948. Paul Herman Müller. gapm.io/xnob.

Novus[1]. Polls for Gapminder in Finland and Norway, April–October 2017. gapm.io/pnovus17a.

Novus[2]. Multiple polls for Gapminder in Sweden, Norway, USA and UK, during the period 2013 to 2017. gapm.io/polls17b.

Novus[3]. Polls for Gapminder in USA and Sweden during April 2017. In USA, November 2013 and September 2016 by GfK Group using KnowledgePanel. In UK, by NatCen. gapm.io/pollnov17bnovus-17b.

Nuclear Notebook. Kristensen, Hans M., and Robert S. Norris. "The Bulletin of the Atomic Scientists' Nuclear Notebook." Federation of American Scientists. https://thebulletin.org/nuclear-notebook-multimedia.

ODI (Overseas Development Institute). Greenhill, Romilly, Paddy Carter, Chris Hoy, and Marcus Manuel. "Financing the future: how international public finance should fund a global social compact to eradicate poverty." ODI, 2015. gapm.io/xodi.

OEC. Simoes, Alexander J. G., and César A. Hidalgo. "The Economic Complexity Observatory: An Analytical Tool for Understanding the Dynamics of Economic Development." Workshops at the Twenty-Fifth AAAI Conference on Artificial Intelligence, 2011. Trade in hs92 category 920.2. String Instruments. gapm.io/xoec17.

The Economic Complexity Observatory. https://atlas.media.mit.edu/en/.

OECD[1] (Organisation for Economic Co-operation and Development). "Why Is Health Spending in the United States So High?" Chart 4: Health spending per capita by category of care, US and selected OECD countries, 2009. Health at a Glance 2011: OECD Indicators. gapm.io/x-ushealth.

OECD[2]. Air and GHG emissions: Carbon dioxide (CO2), Tonnes/capita, 2000–2014. gapm.io/xoecdco2.

OECD[3]. "Indicators of Immigrant Integration 2015". July 2, 2015. OECD, EU gapm.io/xoecdimintegr.

OHDB, Oral Health Database. WHO Collaborating Centre for Education, Training and Research at the Faculty of Odontology, Malmö, Sweden, supported by the WHO Global Oral Health Programme for Oral Health Surveillance and Niigata University, Japan. https://www.mah.se/CAPP/.

Oppenheim Mason, Karen. "Explaining Fertility Transitions." *Demography*, Vol. 34, No. 4, 1997, pp. 443-454. gapm.io/xferttra.

Ostrom, Elinor. *Governing the Commons*. Cambridge, UK: Cambridge University Press, 1990.

OurWorldInData[1]. Roser, Max, and Esteban Ortiz-Ospina. "Declining global poverty: share of people living in extreme poverty, 1820–2015, Global Extreme Poverty." Published online at OurWorldInData.org. Accessed November 20, 2017. ourworldindata.org/extreme-poverty.

OurWorldInData[2]. Roser, Max, and Esteban Ortiz-Ospina "When did literacy start growing in Europe?." Published online at OurWorldInData .org. November 20, 2017. ourworldindata.org/literacy.

OurWorldInData[3]. Roser, Max, and Esteban Ortiz-Ospina. "Child Labor." 2017. Published online at OurWorldInData.org. Accessed November 20, 2017. ourworldindata.org/child-labor.

OurWorldInData[4]. Roser, Max. "Share of World Population Living in Democracies." 2017. Published online at OurWorldInData.org. Accessed November 26, 2017. ourworldindata.org/democracy.

OurWorldInData[5]. Roser, Max. "Ethnographic and Archaeological Evidence on Violent Deaths." Published online at OurWorldInData.org. Accessed November 26, 2017. https://ourworldindata.org/ethnographic -and-archaeological-evidence-on-violent-deaths.

OurWorldInData[6]. Roser, Max, and Mohamed Nagdy. "Nuclear weapons test." 2017. Published online at OurWorldInData.org. Accessed November 14, 2017. https://ourworldindata.org/nuclear-weapons.

OurWorldInData[7]. Number of parties in multilateral environmental agreements based on UNCTAD United Nations Treaty Collection. Published online at OurWorldInData.org. https://ourworldindata.org /grapher/number-of-parties-env-agreements.

OurWorldInData[8]. Tzvetkova, Sandra. "Not All Deaths Are Equal: How Many Deaths Make a Natural Disaster Newsworthy?" July 19, 2017. Published online at OurWorldInData.org. Using results from Eisensee, T., and D. Strömberg. 2007. https://ourworldindata.org/how-many -deaths-make-a-natural-disaster-newsworthy.

OurWorldInData[9]. Ritchie, Hannah and Max Roser. "Energy Production & Changing Energy Sources", Based on Lafond et el. (2017). Published online at OurWorldInData.org. Accessed December 19, 2017. https://ourworldindata.org/energy-production-and-changing-energy-sources/.

OurWorldInData[10]. Roser, Max. "Fertility Rate." Published online at OurWorldInData.org. https://ourworldindata.org/fertility-rate.

Paine, R. R., and J. L. Boldsen. "Linking age-at-death distributions and ancient population dynamics: a case study." 2002. In *Paleodemography: Age distributions from skeletal samples*, ed. R. D. Hoppa and J. W. Vaupel, 169–80. Cambridge, UK: Cambridge University Press.

Paulos, John Allen. *Innumeracy, Mathematical Illiteracy and its Consequences.* New York: Penguin, 1988.

PDNA. Government of Nepal National Planning Commission. *Nepal Earthquake 2015: Post Disaster Needs Assessment*, vol. A. Kathmandu: Government of Nepal, 2015. gapm.io/xnep.

Perry, Mark J. "SAT test results confirm pattern that's persisted for 50 years—high school boys are better at math than girls." *AEIdeas* blog, American Enterprise Institute, September 27, 2016. gapm.io/xsat.

Pew[1]. "Japanese Wary of Nuclear Energy." Pew Research Center Global Attitudes and Trends, June 5, 2012. gapm.io/xpewnuc.

Pew[2]. "Religious Composition by Country, 2010–2050." Pew Research Center Religion & Public Life, April 2, 2015 (table). gapm.io/xpewrel1.

Pew[3]. "The Future of World Religions: Population Growth Projections, 2010–2050." Pew Research Center Religion & Public Life, April 2, 2015. gapm.io/xpewrel2.

Pinker, Steven. *The Better Angels of Our Nature: The Decline of Violence in History and Its Causes.* London: Penguin, 2011.

———. *The Blank Slate: The Modern Denial of Human Nature.* New York: Penguin, 2002.

———. *How the Mind Works.* New York: W.W. Norton, 1997.

———. *The Stuff of Thought.* New York: Viking, 2007.

Platt, John R. "Big News: Wild Tiger Populations Are Increasing for the First Time in a Century." *Scientific American*, April 10, 2016.

PovcalNet "An Online Analysis Tool for Global Poverty Monitoring." Founded by Martin Ravallion, at the World Bank. Accessed November 30, 2017. http://iresearch.worldbank.org/PovcalNet.

PRIO. "The Battle Deaths Dataset version 3.1." Updated in 2006; 1946–2008. See Gleditsch and Lacina (2005), Accessed November 12, 2017. gapm.io/xpriod.

Quétel, Claude. *History of syphilis*. Trans. Judith Braddock and Brian Pike. Cambridge, UK: Polity Press, 1990. gapm.io/xsyph.

Raupach M. R., et al. "Sharing a quota on cumulative carbon emissions." *Nature Climate Change* 4 (2014): 873–79. DOI: 10.1038/nclimate2384. gapm.io/xcar.

Rosling, Hans. "The best stats you've ever seen." Filmed February 2006 in Monterey, CA. TED video, 19:50. https://www.ted.com/talks/hans_rosling_shows_the_best_stats_you_ve_ever_seen. gapm.io/xtedros.

———. "Hans Rosling at World Bank: Open Data." Filmed May 22, 2010, in Washington, DC. World Bank video, 41:54. https://www.youtube.com/watch?v=5OWhcrjxP-E. gapm.io/xwbros.

———. "The magic washing machine." Filmed December 2010 in Washington, DC. TEDWomen video, 9:15. https://www.ted.com/talks/hans_rosling_and_the_magic_washing_machine. gapm.io/tedrosWa.

Rosling, Hans, Yngve Hofvander, and Ulla-Britt Lithell. "Children's death and population growth." *Lancet* 339 (February 8, 1992): 377–78.

Royal Society of London. *Philosophical transactions of the Royal Society of London*. 155 vols. London, 1665–1865. gapm.io/xroys1665.

Sarkees, Meredith Reid, and Frank Wayman. *Resort to War: 1816–2007*. Washington DC: CQ Press, 2010. gapm.io/xcow17.

SCB. System of Environmental and Economic Accounts. gapm.io/xscb2.

Schultz, T. Paul. "Why Governments Should Invest More to Educate Girls." *World Development* 30, no. 2 (2002): 207–25.

SDL. "Slavery in Domestic Legislation", a database by Jean Allain and Dr. Marie Lynch at Queen's University Belfast. http://www.qub.ac.uk/.

Senge, Peter M. *The Fifth Discipline: The Art & Practice of the Learning Organization*. New York: Doubleday, 1990.

Shengmin, Yu, et al. "Study on the Concept of Per Capita Cumulative Emissions and Allocation Options." *Advances in Climate Change Research* 2, no. 2 (June 25, 2011): 79–85. gapm.io/xcli11.

SIPRI Trends in world nuclear forces, 2017. Kile, Shannon N. and Hans M. Kristensen. SIPRI, July 2017. gapm.io/xsipri17.

Smil, Vaclav. *Energy Transitions: Global and National Perspectives*. 2nd ed. Santa Barbara, CA: Praeger, 2016. gapm.io/xsmilen.

———. *Global Catastrophes and Trends: The Next Fifty Years*. Cambridge: MIT Press, 2008. gapm.io/xsmilcat.

Spotify. Web API. https://developer.spotify.com/web-api.

Stockholm Declaration. Fifth Global Meeting of the International Dialogue on Peacebuilding and Statebuilding, 2015. https://www.pbsbdialogue .org/en.

Sundberg, Ralph and Erik Melander. "Introducing the UCDP Georeferenced Event Dataset", Journal of Peace Research, vol. 50, no. 4, 523-532.

Sundin, Jan, Christer Hogstedt, Jakob Lindberg, and Henrik Moberg. *Svenska folkets hälsa i historiskt perspektiv.* Barnhälsans historia, page 122. Solna: Statens folkhälsoinstitut, 2005. gapm.io/xsfhi5.

Tanigawa, Koichi, et al. "Loss of life after evacuation: lessons learned from the Fukushima accident." *Lancet* 379, no. 9819 (March 10, 2012): 889–91. gapm.io/xfuk.

Tavris, Carol, and Elliot Aronson. *Mistakes Were Made (But Not by Me): Why We Justify Foolish Beliefs, Bad Decisions, and Hurtful Acts.* New York: Harcourt, 2007.

Tetlock, P.E., and D. Gardner. *Superforecasting: The Art and Science of Prediction.* New York: Crown, 2015.

The Economist[1]. "The tragedy of the high seas." *Economist*, February 22, 2014. gapm.io/xeconsea.

The Economist[2]. "Democracy Index from the Economist Intelligence Unit." Accessed December 2, 2017. gapm.io/xecodemi.

Tylleskär, Thorkild. "KONZO—the walk of the chameleon." Video, a group work in global nutrition, featuring Dr. Jean-Pierre Banea-Mayambu (head of Pronanut), Dr. Desire Tshala-Katumbay (from the neurology clinic at Centre Neuropsychopathologique, CNPP, Kinshasa), and students in nutrition at Uppsala University, Sweden, 1995. gapm.io /xvkonzo.

UCDP[1] (Uppsala Conflict Data Program). Battle-Related Deaths Dataset, 1989 to 2016, dyadic, version 17.1. See Allansson et al., dyadic, version 17.1. http://ucdp.uu.se/downloads.

UCDP[2]. Uppsala Conflict Data Program, Georeferenced Event Dataset (GED) Global version 17.1 (2016), See Sundberg et al (2013). Department of Peace and Conflict Research, Uppsala University, http://ucdp .uu.se/downloads.

UN Comtrade. https://comtrade.un.org/.

UN Statistic Division. "Developing regions". Accessed December 20, 2017. gapm.io/xunsdef.

UN-IGME (United Nations Inter-agency Group for Child Mortality Estimation). "Child Mortality Estimates." Last modified October 19, 2017. http://www.childmortality.org.

UN-Pop[1] (UN Population Division). Population, medium fertility variant. World Population Prospects 2017. United Nations, Department of Economic and Social Affairs, Population Division. https://esa.un.org/unpd/wpp.

UN-Pop[2]. Annual age composition of world population, medium fertility variant. World Population Prospects 2017. UN Population Division. https://esa.un.org/unpd/wpp.

UN-Pop[3]. Indicators: Life expectancy and total fertility rate (medium fertility variant). World Population Prospects 2017. UN Population Division. Accessed September 2, 2017. https://esa.un.org/unpd/wpp.

UN-Pop[4]. Annual population by age—Female, medium fertility variant. World Population Prospects 2017. UN Population Division. Accessed November 7, 2017. gapm.io/xpopage.

UN-Pop[5]. World Population Probabilistic Projections. Accessed November 29, 2017. gapm.io/xpopproj.

UN-Pop[6]. "The impact of population momentum on future population growth." *Population Facts* no. 2017/4 (October, 2017): 1–2. gapm.io/xpopfut.

UN-Pop[7]. Andreev, K., V. Kantorová, and J. Bongaarts. "Demographic components of future population growth." Technical paper no. 2013/3. United Nations DESA Population Division, 2013. gapm.io/xpopfut2.

UN-Pop[8]. Deaths (both sexes combined), medium fertility variant. World Population Prospects 2017. UN Population Division. Accessed December 2, 2017. gapm.io/xpopdeath.

UN-Pop[9]. World Contraceptive Use 2017. World Population Prospects 2017. UN Population Division, March 2017. Accessed December 2, 2017. gapm.io/xcontr.

UNAIDS. "AIDSinfo." Accessed October 4, 2017. http://aidsinfo.unaids.org.

UNDESA (United Nations Department of Economic and Social Affairs). "Electricity and education: The benefits, barriers, and recommendations for achieving the electrification of primary and secondary schools." December 2014. gapm.io/xdessel.

UNEP[1] (United Nations Environment Programme). *Towards a Pollution-Free Planet*. Nairobi: United Nations Environment Programme, 2017. gapm.io/xpolfr17.

UNEP[2]. Regional Lead Matrix documents published between 1990 and 2012. gapm.io/xuneplmats.

UNEP[3]. "Leaded Petrol Phase-out: Global Status as at March 2017." Accessed November 29, 2017. gapm.io/xunepppo.

UNEP[4]. Ozone data access center: ODS consumption in ODP tonnes. Data updated November 13, 2017. Accessed November 24, 2017. gapm .io/xods17.

UNEP[5]. The World Database on Protected Areas (WDPA). UNEP, IUCN, and UNEP-WCMC. https://protectedplanet.net.

UNEP[6]. Protected Planet Report 2016. UNEP-WCMC and IUCN, Cambridge, UK, and Gland, Switzerland, 2016. Accessed December 17, 2017. gapm.io/xprotp16.

UNESCO[1] (United Nations Educational, Scientific and Cultural Organization). "Education: Completion rate for primary education (household survey data)." Accessed November 5, 2017. gapm.io/xcomplr.

UNESCO[2]. "Education: Literacy rate." Last modified July 2017. Accessed November 5, 2017. gapm.io/xuislit.

UNESCO[3]. "Education: Out-of-school rate for children of primary school age, female." Accessed November 26, 2017. gapm.io/xuisoutsf.

UNESCO[4]. "Rate of out-of-school children." Accessed November 29, 2017. gapm.io/xoos.

UNESCO[5]. "Reducing global poverty through universal primary and secondary education." June 2017. gapm.io/xprimsecpov.

UNFPA[1] (United Nations Population Fund). "Sexual & reproductive health." Last updated November 16, 2017. http://www.unfpa.org/sexual -reproductive-health.

UNHCR (United Nations High Commissioner for Refugees). "Convention and protocol relating to the status of refugees." UN Refugee Agency, Geneva. gapm.io/xunhcr.

UNICEF-MICS. Multiple Indicator Cluster Surveys. Funded by the United Nations Children's Fund. Accessed November 29, 2017. http:// mics.unicef.org.

UNICEF[1]. *The State of the World's Children 1995*. Oxford, UK: Oxford University Press, 1995. gapm.io/xstchi.

UNICEF[2]. "Narrowing the Gaps—The Power of Investing in the Poorest Children." July 2017. gapm.io/xunicef2.

UNICEF[3]. "Diarrhoea remains a leading killer of young children, despite the availability of a simple treatment solution." Accessed September 11, 2017. gapm.io/xunicef3.

UNICEF[4]. "The State of the World's Children 2013—Children with Disabilities." 2013. gapm.io/x-unicef4.

UNICEF[5]. "Vaccine Procurement Services". https://www.unicef.org/supply/index_54052.html.

UNISDR (United Nations Office for Disaster Risk Reduction). "Heat wave in Europe in 2003: new data shows Italy as the most affected country." UNISDR, 2003. gapm.io/x-unicefC5.

US Census Bureau. Current Population Survey, 2017 Annual Social and Economic Supplement. Table: "FINC01_01. Selected Characteristics of Families by Total Money Income in: 2016," monetary income, all races, all families. gapm.io/xuscb17.

US-CPS. Current Population Survey 2016: Family Income in 2016.

USAID-DHS[1]. Demographic and Health Surveys (DHS), funded by USAID. https://dhsprogram.com.

USAID-DHS[2]. Bietsch, Kristin, and Charles F. Westoff., *Religion and Reproductive Behavior in Sub-Saharan Africa*. DHS Analytical Studies No. 48. Rockville, MD: ICF International, 2015. gapm.io/xdhsarel.

van Zanden[1]. van Zanden, Jan Luiten, Joerg Baten, Peter Foldvari, and Bas van Leeuwen. "World Income Inequality: The Changing Shape of Global Inequality 1820–2000." Utrecht University, 2014. http://www.basvanleeuwen.net/bestanden/WorldIncomeInequality.pdf.

van Zanden[2]. van Zanden, Jan Luiten, and Eltjo Buringh. "Rise of the West: Manuscripts and Printed Books in Europe: A long-term perspective from the sixth through eighteenth centuries." *Journal of Economic History* 69, no. 2 (February 2009): 409–45. gapm.io/xriwe.

van Zanden[3], van Zanden, Jan Luiten, et al., eds. *How Was Life? Global Well-Being Since 1820*. Paris: OECD Publishing, 2014. gapm.io/x-zanoecd.

WEF (World Economic Forum). "Davos 2015—Sustainable Development: Demystifying the Facts." Filmed Davos, Switzerland, January 2015. WEF video, 15:42. Link to 5 minutes 18 seconds into the presentation, when Hans show the audience results: https://youtu.be/3pVlaEbpJ7k?t=5m18s.

White[1]. White, Matthew. *The Great Big Book of Horrible Things*. New York: W.W. Norton, 2011.

White[2]. White, Matthew. Estimates of death tolls in World War II. Necrometrics. http://necrometrics.com/20c5m.htm#Second.

WHO[1]. "Global Health Observatory data repository: Immunization." Accessed November 2, 2017. gapm.io/xwhoim.

WHO[2]. Safe abortion: Technical & policy guidance for health systems. gapm.io/xabor.

WHO[3]. WHO Ebola Response Team. "Ebola Virus Disease in West Africa—The First 9 Months of the Epidemic and Forward Projections." *New England Journal of Medicine* 371 (October 6, 2014): 1481–95. gapm.io/xeboresp.

WHO[4]. "Causes of child mortality." Global Health Observatory (GHO) data. Accessed September 12, 2017. gapm.io/xeboresp2.

WHO[5]. "1986–2016: Chernobyl at 30." April 25, 2016. gapm.io/xwhoc30.

WHO[6]. "The use of DDT in malaria vector control: WHO position statement." Global Malaria Programme, World Health Organization, 2011. gapm.io/xwhoddt1.

WHO[7]. "DDT in Indoor Residual Spraying: Human Health Aspects— Environmental Health Criteria 241." World Health Organization, 2011. gapm.io/xwhoddt2.

WHO[8]. "WHO Global Health Workforce Statistics." World Health Organization, 2016. gapm.io/xwhowf.

WHO[9]. Situation updates—Pandemic. gapm.io/xwhopand.

WHO[10]. Data Tuberculosis (TB) Global Health Observatory (GHO) data, http://www.who.int/gho/tb/.

WHO[11]. "What is multidrug-resistant tuberculosis (MDR-TB) and how do we control it?" gapm.io/xmdrtb.

WHO[12]. "Global Health Expenditure Database." Last updated December 5, 2017. http://apps.who.int/nha/database.

WHO[13]. Ebola situation reports. gapm.io/xebolawho.

WHO[14]. Antimicrobial resistance. gapm.io/xantimicres.

WHO[15]. Neglected tropical diseases. gapm.io/xnegtrop.

WHO[16]. "Evaluation of the international drinking water supply and sanitation decade, 1981-1990", World Health Organization, November 21, 1991. Executive board, eighty-ninth session. Page 4. gapm.io/xwhow90.

WHO[17]. Emergencies preparedness, response. Situation updates - Pandemic (H1N1) 2009 http://www.who.int/csr/disease/swineflu/updates/en/index.html.

WHO[18]. Data Tuberculosis (TB) Global Health Observatory (GHO) data, http://www.who.int/gho/tb/en/.

WHO/UNICEF. "Ending Preventable Child Deaths from Pneumonia and Diarrhoea by 2025." World Health Organization/The United Nations Children's Fund (UNICEF), 2013. gapm.io/xpneuDiarr.

WHO/UNICEF JMP (Joint Monitoring Programme). "Drinking water, sanitation and hygiene levels," 2015. https://washdata.org/data.

Wikipedia[1]. "Timeline of abolition of slavery and serfdom." https://en.wikipedia.org/wiki/Timeline_of_abolition_of_slavery_and_serfdom.

Wikipedia[2]. "Capital punishment by country: Abolition chronology." https://en.wikipedia.org/wiki/Capital_punishment_by_country#Abolition_chronology.

Wikipedia[3]. "Feature film: History." https://en.wikipedia.org/wiki/Feature_film#History.

Wikipedia[4]. "Women's suffrage." https://en.wikipedia.org/wiki/Women%27s_suffrage.

Wikipedia[5]. "Sound recording and reproduction: Phonoautograph." https://en.wikipedia.org/wiki/Sound_recording_and_reproduction#Phonautograph.

Wikipedia[6]. "World War II casualties." https://en.wikipedia.org/wiki/World_War_II_casualties.

Wikipedia[7]. "List of terrorist incidents: 1970–present." https://en.wikipedia.org/wiki/List_of_terrorist_incidents#1970–present.

Wikipedia[8]. "Cobratoxin: Multiple sclerosis." https://en.wikipedia.org/wiki/Cobratoxin#cite_note-pmid21999367-8.

Wikipedia[9]. "Charles Waterton." https://en.wikipedia.org/wiki/Charles_Waterton.

Wikipedia[10]. "Recovery position." https://en.wikipedia.org/wiki/Recovery_position.

World Bank[1]. "Indicator GDP per capita, PPP (constant 2011 international $)." International Comparison Program database. Downloaded October 22, 2017. gapm.io/xwb171.

World Bank[2]. "World Bank Country and Lending Groups." Accessed November 6, 2017. gapm.io/xwb172.

World Bank[3]. "Primary completion rate, female (% of relevant age group)." Accessed November 5, 2017. gapm.io/xwb173.

World Bank[4]. "Population of Country Income Groups in 2015—Population, total." Accessed November 7, 2017. gapm.io/xwb174.

World Bank[5]. "Poverty headcount ratio at $1.90 a day (2011 PPP) (% of population)." Development Research Group. Downloaded October 30, 2017. gapm.io/xwb175.

World Bank[6]. "Indicator Access to electricity (% of population)." Sustainable Energy for All (SEforALL) Global Tracking Framework. International Energy Agency and the Energy Sector Management Assistance Program, 2017. gapm.io/xwb176.

World Bank[7]. "Life expectancy at birth, total (years)." United Nations Statistical Division. Population and Vital Statistics Reports (various years). Accessed November 8, 2017. gapm.io/xwb177.

World Bank[8]. "Improved water source (% of population with access)." WHO/UNICEF Joint Monitoring Programme (JMP) for Water Supply and Sanitation. Accessed November 8, 2017. gapm.io/xwb178.

World Bank[9]. "Immunization, measles (% of population with access)." Accessed November 8, 2017. gapm.io/xwb179.

World Bank[10]. "Prevalence of undernourishment (% of population)." Food and Agriculture Organisation. Accessed November 8, 2017. gapm.io/xwb1710.

World Bank[11]. "Out-of-pocket health expenditure (% of total expenditure on health)." Global Health Expenditure database, 2017. gapm.io/xwb1711.

World Bank[12]. Narayan, Deepa, Raj Patel, et al. *Voices of the Poor: Can Anyone Hear Us?* New York: Oxford University Press, 2000. gapm.io /xwb1712.

World Bank[13]. "International tourism: number of departures." Yearbook of Tourism Statistics, Compendium of Tourism Statistics and data files, World Tourism Organization, 2017. gapm.io/xwb1713.

World Bank[14]. "Beyond Open Data: A New Challenge from Hans Rosling." Live GMT, June 8, 2015. gapm.io/xwb1714.

World Bank[15]. Khokhar, Tariq. "Should we continue to use the term 'developing world'?" *The Data* blog, World Bank, November 16 , 2015. gapm.io/xwb1715.

World Bank[16]. "Income share held by highest 10%." Development Research Group, 2017. gapm.io/xwb1716.

World Bank[17]. Jolliffe, Dean, and Espen Beer Prydz. "Estimating International Poverty Lines from Comparable National Thresholds." World Bank Group, 2016. gapm.io/xwb1717.

World Bank[18]. "Mobile cellular subscriptions." International Telecommunication Union, World Telecommunication/ICT Development Report and database. Downloaded November 26, 2017. gapm.io/xwb1718.

World Bank[19]. "Individuals using the Internet (% of population)." International Telecommunication Union, World Telecommunication/ICT Development Report and database. Downloaded November 27, 2017. gapm.io/xwb1719.

World Bank[20] Global Consumption Database. http://datatopics.worldbank.org/consumption.

World Bank[21]. "School enrollment, primary and secondary (gross), gender parity index (GPI)." United Nations Educational, Scientific, and Cultural Organization (UNESCO) Institute for Statistics, 2017. gapm.io/xwb1721.

World Bank[22]. "Global Consumption Database." World Bank Group, 2017. gapm.io/xwb1722.

World Bank[23]. "Physicians (per 1,000 people)." Selected countries and economies; Sweden and Mozambique. World Health Organization, Global Health Workforce Statistics, OECD, 2017. gapm.io/xwb1723.

World Bank[24]. "Health expenditure per capita, PPP (constant 2011 international $)." World Health Organization Global Health Expenditure Database, 2017. gapm.io/xwb1724.

World Bank[25]. Newhouse, David, Pablo Suarez-Becerra, and Martin C. Evans. "New Estimates of Extreme Poverty for Children—Poverty and Shared Prosperity Report 2016: Taking On Equality." Policy Research Working Paper no. 7845. World Bank, Washington, DC, 2016.

World Bank[26] Group, Poverty and Equality Global Practice Group, October 2016. gapm.io/xwb1726.

World Bank[27]. World Bank Open Data platform. https://data.worldbank.org.

WorldPop. Case Studies—Poverty. gapm.io/xworpopcs.

WWF. Tiger—Facts. 2017. Accessed November 5, 2017. gapm.io/xwwf
tiger.

YouGov[1]. November–December 2015. Poll results: gapm.io/xyougov15.

YouGov[2]. Poll about fears. 2014. gapm.io/xyougov15.

Zakaria, Fareed. *The Future of Freedom: Illiberal Democracy at Home and Abroad.* New York: W.W. Norton, 2003.

———. *The Post-American World.* New York: W.W. Norton, 2008.

BIOGRAPHICAL NOTES

Hans Rosling

Hans was born in Uppsala, Sweden, in 1948. He studied statistics and medicine at Uppsala University and public health at St. John's Medical College, Bangalore, India, qualifying as a doctor in 1976. Between 1974 and 1984 he was home full time caring for his three children for a total of 18 months. From 1979 to 1981, Hans was district medical officer in Nacala, Mozambique, where he discovered a previously unrecognized paralytic disease now known as konzo. His subsequent investigations into this disease earned him a PhD from Uppsala University in 1986. From 1997, Hans was professor of International Health at Karolinska Institutet, the medical university in Stockholm, Sweden. His research focused on the links between economic development, agriculture, poverty, and health. He also started new courses, launched research partnerships, and co-authored a textbook on Global Health.

In 2005, Hans co-founded the Gapminder Foundation with his son Ola Rosling and daughter-in-law Anna Rosling Rönnlund, with a mission to fight devastating ignorance with a fact-based worldview that everyone can understand. Hans has lectured to financial institutions, corporations, and nongovernmental organizations and his ten TED talks have been viewed over 35 million times.

Hans was an adviser to the World Health Organization, UNICEF, and several aid agencies, and cofounded Médecins Sans Frontières in Sweden. He has developed and presented three BBC documentaries: *The Joy of Stats* in 2010, *Don't Panic—The Truth About Population* in 2013, and *Don't Panic: How to End Poverty in 15 Years* in 2015. Hans was a member of the International Group of the Swedish Academy of Science and of the Global Agenda Network of the World Economic Forum in Switzerland. He was listed as one of the 100 leading global thinkers by *Foreign Policy* magazine in 2009, as one of the 100 most creative people in business by *Fast Company* magazine in 2011, and as one of *Time* magazine's 100 most influential people in the world in 2012.

Hans and his wife, Agneta Rosling, have three children: Anna, Ola, and Magnus. Hans died on February 7, 2017.

Ola Rosling

Ola Rosling was born in Hudiksvall, Sweden, in 1975. He is a cofounder of the Gapminder Foundation and was its director from 2005 to 2007 and from 2010 to the present day.

Ola invented and developed Gapminder's ignorance tests, its structured ignorance measuring project, and its certification process. He crunched the data and developed the materials for most of Hans's TED talks and lectures. From 1999, Ola led the development of the famous animated bubble-chart tool called Trendalyzer, used by millions of students across the world to understand multidimensional time series. In 2007, the tool was acquired by Google, where Ola led the Google Public Data Team between 2007 and 2010. He then returned to Gapminder to develop new free teaching materials.

Ola lectures widely and his joint TED talk with Hans has been viewed millions of times.

Ola has received several awards for his work at Gapminder, including a Résumé Super-communicator Award and the Guldägget Titanpriset in 2017 and the Niras International Integrated Development Prize in 2016.

Ola is married to Anna Rosling Rönnlund. They have three children: Max, Ted, and Ebba.

Anna Rosling Rönnlund

Anna was born in Falun, Sweden, in 1975. She holds degrees in sociology from Lund University and photography from Gothenburg University and is a cofounder and vice president of the Gapminder Foundation.

Anna is a lecturer and the guardian of the end user at Gapminder, making sure that everything Gapminder does is easy to understand. Together with Ola, Anna directed Hans's TED talks and other lectures, developed the Gapminder graphics and slides, and designed the user interface of the animated bubble-chart tool Trendalyzer. When the tool was acquired by Google in 2007, she went to work for Google as a senior usability designer. In 2010, Anna returned to Gapminder to develop new free teaching materials.

Dollar Street, launched in 2016, is Anna's brainchild and the subject of her 2017 TED talk.

Anna has won several awards for her work at Gapminder, including a Résumé Super-communicator Award, the Guldägget Titanpriset, and the Fast Company World Changing Ideas Award in 2017.

Anna is married to Ola Rosling. They have three children: Max, Ted, and Ebba.

INDEX